Mistletoe: From Mythology to Evidence-Based Medicine

Translational Research in Biomedicine

Vol. 4

Series Editor

Samuel H.H. Chan Kaohsiung

Associate Editor

Julie Y.H. Chan Kaohsiung

The Chang Gung Medical Foundation is the patron of this book series.

Mistletoe: From Mythology to Evidence-Based Medicine

Volume Editors

Kurt S. Zänker Witten
Srini V. Kaveri Paris/Mumbai

5 figures, 5 in color, 5 tables, 2015

KARGER

Basel · Freiburg · Paris · London · New York · Chennai · New Delhi ·
Bangkok · Beijing · Shanghai · Tokyo · Kuala Lumpur · Singapore · Sydney

Kurt S. Zänker
Fakultät für Gesundheit (Department für
Humanmedizin)
Institut für Immunologie
Universität Witten/Herdecke
Stockumer Straße 10
D–58453 Witten (Germany)

Srini V. Kaveri
Centre de Recherche des Cordeliers
15 rue de l'Ecole de Médecine
F–75006 Paris (France) and
International Associated Laboratory IMPACT
Indian council of Medical Research
IN–400012 Mumbai (India)

Library of Congress Cataloging-in-Publication Data

Mistletoe (Zänker)
  Mistletoe : from mythology to evidence-based medicine / volume editors,
Kurt S. Zänker, Srini V. Kaveri.
     p. ; cm. -- (Translational research in biomedicine, ISSN 1662-405X ;
v. 4)
  Includes bibliographical references and indexes.
  ISBN 978-3-318-05444-6 (hard cover : alk. paper) -- ISBN 978-3-318-05445-3
(e-ISBN)
  I. Zänker, Kurt S., editor. II. Kaveri, Srini V., editor. III. Title. IV.
Series: Translational research in biomedicine ; v. 4. 1662-405X
  [DNLM: 1. Antineoplastic Agents, Phytogenic--therapeutic use. 2. Viscum
album. 3. Anti-Inflammatory Agents--therapeutic use. 4. Fatigue--drug
therapy. QZ 267]
  RS431.A64
  615.7'98--dc23
                                                                    2015008686

© Copyright 2015 by S. Karger AG, P.O. Box, CH–4009 Basel (Switzerland)
www.karger.com
Printed in Germany on acid-free and non-aging paper (ISO 9706) by Kraft Druck GmbH, Ettlingen
ISSN 1662–405X
e-ISSN 1662–4068
ISBN 978-3–318–05444–6
e-ISBN 978-3–318–05445–3

Contents

Foreword

Welcome to Volume Four of *Translational Research in Biomedicine*, a monograph series dedicated to the dissemination of seminal information in contemporary biomedicine with a translational orientation. This volume marks the second installment of this series under the generous patronage of Chang Gung Medical Foundation of Taiwan. This patronage substantially reduces the increasing financial constraints on scientific publication and allows us to concentrate on publishing text on timely and crucial themes in translational medicine.

This volume is designed to be a key reference on mistletoe. It is intended to reveal all of the conflicts of interest, bias, omissions, and scientific censorship related to the use of mistletoe in oncology. In the spirit of translational medicine, this volume will provide readers with an understanding of the mythology, the culture, the contextuality and the personalized treatment associated with mistletoe, along with fundamental knowledge based on molecular biology, chemistry and biochemistry. Whereas the majority of the literature on plant and natural products has a preferential Eastern flavor, this volume is unique in its own right because it represents our contemporary understanding of mistletoe from a European perspective.

I wish to express my deepest appreciation to Professors Kurt S. Zänker and Srini V. Kaveri, whose patience and sustaining efforts have made this timely volume, or 'Mistletoe: From Mythology to Evidence-Based Medicine', a reality. I also wish to acknowledge the capable hands of Thomas Nold and Ricarda Cueni at S. Karger AG during the development and production of this volume. Last but not least, the publication of *Translational Research in Biomedicine* would not have been possible without the foresight, enthusiasm and whole-hearted support of my dear friend Dr. Thomas Karger.

Samuel H.H. Chan
Series Editor

Preface

It is a great honor for us (Kurt S. Zänker (Witten) and Srini V. Kaveri (Paris/Mumbai)) to edit this book on mistletoe, *From Mythology to Evidence-Based Medicine*, within the book series *Translational Research in Biomedicine*, published by Karger, as guest editors. With his outstanding reputation in biomedicine and the peer-reviewed approval of the chapters' content, the Series Editor – Samuel H.H. Chan (Kaohsiung) – is a guarantor of a high international scientific standard for revealing and disseminating experimental and clinical results, embedded into a cultural context, regarding the use of mistletoe preparations in biomedicine.

We have been fortunate to recruit competent scientists who have been working in the field of translational mistletoe research for decades, and we thank them for their distinguished contributions.

Thomas Kuhn [1] fathered, defined and popularized the concept of 'paradigm shift'. Currently, a long-standing, historically based view of medicine centered on the intuition, passion and creativity of the individual scientist is being replaced. The replacement is taking place stepwise via a series of peaceful interludes punctuated by biotechnologically and digitally violent revolutions that place value on data collection itself in biomedicine. The 21st century is predestined for a paradigm shift in biomedicine because exciting molecular biology results in biomedical research, the dramatic explosion of (nano-)biotechnology and the digital revolution provide the basis for P4 Medicine in cancer research [2]. A mandatory prerequisite in personalized medicine – and one of the P4 items – is to collect the greatest quantity of disease-related data sets possible in the hope of better diagnosis and treatment. However, where are the sophisticated algorithms for mining; integrating; modeling; and, at the patient's end, dissecting the large, heterogeneous biological data sets to generate an individualized, actionable and meaningful model to determine what is the best individual therapy at the right time?

At this time, biomedicine is addressing two ideas: i) the idea of applying an object (disease)-orientated collection of computing-powered data sets – from bench to bytes to bedside – and ii) the conservative idea of converting what is perceived into a personal scientific concept to develop hypothesis-driven biomedical starting points for perceiving and reconstructing a patient's individual world for better diagnosis and therapy.

It might sound complex, but behind these two ideas lie some simple concepts. In the future, for successful biomedicine, we need both, and both should provide mutual assistance in order to facilitate biomedical applications of their directives to the well-being of patients.

Biomedicine is not only basic science but also humanities and cultural sciences. The history of medicine clearly shows that many fundamental discoveries and inventions in medicine are the results of the interactions of these entities.

This book is a treasure chest for all physicians and caregivers in biomedicine who want to see how mistletoe research developed into its translational and clinical applications today. For the first time, this book depicts the trajectory of mistletoe research in oncology from the beginning, following the philosophy – from mythology to evidence-based medicine – behind plotting the scientific course of human endeavors in a digital world.

The book starts with a chapter by *Hartmut Ramm* on the ancient use of mistletoe and its cultural myths in religion, in health and in disease. The molecular diversity of mistletoe constituents, as analyzed by modern chemical methods, is described by *Konrad Urech* and *Stephan Baumgartner*. The biodiversity of these substances is set in a functional context of anti-cancer activities by *Henning Schramm*. The chapters reporting on translational and clinical mistletoe research start with a chapter by *Udo Schumacher* and *Uwe Pfüller*, who give a timeline of important stages and results in the development of mistletoe research. *Sonja Schötterle* and *Ulrike Naumann* provide evidence that mistletoe-based drugs are candidates for use as concomitant therapeutics in parallel to standard therapy, e.g. for glioblastoma. *Danijel Galun, Wilfried Tröger* and *Miroslav Milicevic* give a summary of the use of mistletoe preparations as supportive care in cancer surgery, with special reference to pancreatic cancer. *Sri Ramula Elluru, Chaitrali Saha, Pushpa Hedge et al.* present translational results obtained for mistletoe preparations targeting cellular and molecular mechanisms of the immune system and inflammation. These translational research data provide a rationale for addressing cancer-related fatigue (CRF) clinically using mistletoe. It is very likely that inflammatory processes are involved in the etiology of CRF. Therefore, plant-derived anti-inflammatory substances, e.g. those derived from mistletoe preparations, might be the first candidates for fighting against CRF when applied together with conventional anti-tumor protocols. *Kurt S. Zänker* gives clinical evidence that CRF can be alleviated by the use of mistletoe preparations in oncology.

We are grateful to the publisher Karger (Switzerland) and to Samuel H.H. Chan (Kaohsiung), the Series Editor, for this platform to communicate a better understanding of the medical history; the plant chemistry; and the experimental, translational and functional activities of mistletoe constituents in the field of oncology to the appropriate scientific communities. Taken together, this information contributes to holistic and personalized medicine.

Winter 2014

Kurt S. Zänker, Witten

Srini V. Kaveri, Paris/Mumbai

Volume Editors

References

1 Kuhn T: The Structure of Scientific Revolution. Chicago, University of Chicago Press, 1962.

2 Hood L, Friend SH: Predictive, personalized, preventive, participatory (P4) cancer medicine. Nat Rev Clin Oncol 2011;8:184–187.

Zänker KS, Kaveri SV (eds): Mistletoe: From Mythology to Evidence-Based Medicine.
Transl Res Biomed. Basel, Karger, 2015, vol 4, pp 1–10 (DOI: 10.1159/000375421)

Mistletoe through Cultural and Medical History: The All-Healing Plant Proves to Be a Cancer-Specific Remedy

Hartmut Ramm

Hiscia Institute, Society for Cancer Research, Arlesheim, Switzerland

Abstract

For millennia, mistletoe *(Viscum album)* has been an important element of human culture. Its uses have ranged from practical, for example, as animal food, to ceremonial, for example, in initiation rituals; however, special attention has always been given to its potential as a medicinal plant. From the time of Hippocrates, physicians have recommended mistletoe to cure different diseases, especially epilepsy. Juice or powder gained from leaves, berries, or stems of mistletoe were given as a drink or applied as a plaster or ointment. Often, mistletoe was mixed with other organic or inorganic ingredients. New interest arose in the early 20th century, when mistletoe's potential in cancer treatment was determined. While Rudolf Steiner developed the basic concept, Ita Wegman was the first physician to specifically apply a mistletoe extract in cancer patients. Steiner based his concept on careful observation of morphological patterns, considered polar qualities of mistletoe in summer and winter, and designed a unique pharmaceutical processing. From ancient times, oak mistletoe was regarded as a most valuable, but very rare, medicinal plant, and it has been used in cancer therapy since 1927. Due to the limited availability of oak mistletoe, a programme was launched to safeguard the stocks of oak mistletoe by searching for wild sites in France and by cultivating *V. album* on *Quercus robur* and *Q. petraea*. One century after mistletoe was introduced into cancer therapy, results from more than 100 clinical studies have indicated that mistletoe has several beneficial effects. Recently, new clinical studies have confirmed the specific potential of oak mistletoe. Once revered as an important healing plant by Celtic druids, mistletoe is now proving its relevance as a cancer-specific remedy more and more.

Archaeological Findings and Early Scientific Observations

The white-berried mistletoe (*Viscum album* L.) has been an element of human culture for several thousand years [1]. Archaeological findings on Neolithic sites indicate that this evergreen tree parasite has been used to feed animals, especially during long winter periods. Based on samples of mistletoe stems, leaves and berries from different locations, archaeologists assume that these early human civilisations also used mistletoe for ritual and medicinal purposes [2].

The first scientific observations were documented by Theophrastus (371–287 BC) and were related to the specific biology of mistletoe as a parasitic plant. According to Tubeuf [1], the Greek author differentiated between the types of mistletoe that could be found on different host trees. He compared its mode of establishment on trees with grafting but had already remarked that birds were essential for the dispersal of mistletoe seeds. Four centuries later, the Roman author Pliny the Elder (23–79 AD) referred to this early scientific documentation of mistletoe when he compiled his *Naturalis Historia* [3].

Mythological Aspects

In his *Aeneid* [4], the well renowned Roman poet and philosopher Virgil (70–19 BC) documented the mystic value that mistletoe had in the ancient Greek and Roman periods. In the sixth book of his epic, when the hero Aeneas is striving to enter the underworld, a Sybil tells him that he must find the golden bough that, like the *Viscum*, is twinkling in the bare crown of an oak. Only then will he be able to pass the initiation without harm. Two doves then guide him to the hidden place where the hero can break the branch that will protect him on his dangerous path. Finally, after sacrificing the golden bough to the goddess of the underworld, Aeneas gains access to higher levels of knowledge [5].

One hundred years later, Pliny mentioned the high regard the Gallic druids had for the mistletoe in his *Naturalis Historia*. The Celtic culture had been predominant in central and western parts of Europe since 800 BC, but it began to decline after the Roman army defeated the Gallic troops around 50 BC. Because the Celtic people did not preserve any of their cultural values in written form, Pliny's report saved mistletoe from oblivion, describing their vision of it as *omnia sanans* [all-healing]. He describes how a white-dressed druid would harvest the mistletoe with a golden sickle on the sixth day after the new moon and how the branches were caught in a white cloak so that they would not touch the ground [3].

Oral tradition preserved in the Austrian Alps confirmed that a ritual mistletoe harvest was also performed by Celtic druids in central Europe and provided more details of the ceremony [6]. While Pliny had mentioned only one druid, the report from Styria, the former Roman province of Noricum, described two druids with polar attributes who performed the ceremonious mistletoe harvest together: the white-dressed druid riding on a white horse with a golden bridle and using a golden sickle matches Pliny's white-dressed druid. The other druid rode a black horse with a silver bridle and wore a black coat but had no working tool and did not actively join the harvest process [5].

Polarity was also factor for Snorri Sturluson (1179–1241 AD), a scholar from Iceland who provided the Nordic perspective on mistletoe in the *Younger Edda* [7]. This myth, written in prose around 1220, included the story of the beloved god Balder, who dreamed that his life was in danger. His mother extracted oaths from all beings to not harm her son, overlooking only one – the mistletoe. The cunning Loke tricked her into telling him this, and without hesitation, Loke ran to find the mistletoe and then convinced blind Hoder to use the branch on Balder, who was killed by this joint attack. The envy of the bright and active Loke and the raw power of the passive and blind Hoder reflect the polarity of forces that caused Balder's death when his opponents used the mistletoe branch as a weapon [5].

It is worth mentioning that in the Romantic period, Vincenzo Bellini (1801–1835) revived the cultic adoration of mistletoe by Celtic druids in his famous opera *Norma*. The high priestess of the druids is in love with the Roman proconsul in Gaul and pleads for peace on the altar of the chaste goddess with an oak mistletoe branch in her hand.

Mistletoe in the History of Medicine

Early reports on the medical use of mistletoe originate from the Greek physician Dioscorides (40–90 AD). However, Hippocrates (460–377 BC) and his students had already used mistletoe to treat diseases of the spleen and complaints associated with menstruation [1].

In the fifth book of *De Medicina*, the Platonist Celsus (25 BC-50 AD) describes mistletoe as an ingredient of several formulations. Mixed with other organic or inorganic substances in plasters or emollients, mistletoe was used to reduce pain or swelling, for suppuration or to treat abscesses, carcinoids, or scrofula [8]. Pliny also mentioned mistletoe as beneficial for treating epilepsy, infertility and ulcers [3].

The German abbess Hildegard von Bingen (1098–1179) recommended mistletoe for the treatment of spleen and liver diseases, and Paracelsus (Theophrastus von Hohenheim, 1493–1541) gave detailed instructions on using oak mistletoe to cure epilepsy. European herbal books in the 16th century describe mistletoe as warming, softening and an astringent. They also attribute specific medicinal value to oak mistletoe for treating epilepsy as well as diseases of the kidney and spleen. Furthermore, they mention the general use of mistletoe, together with other plants, as poultices and plasters for treating ulcers, bone-fractures and labour-pain [9].

At the dawn of the scientific age, British chemist Sir Robert Doyle (1627–1691), the founding member of the Royal Society, recommended drinking pulverised mistletoe dissolved in black cherry juice for several days around the full moon as a remedy for epilepsy [10]. A German encyclopaedia compiled in the middle of the 18th century specifies several ways to process remedies from oak mistletoe for leprosy, epistaxis, round worms, and other diseases [11]. In addition, in 1720, the English physician Colbatch recommended mistletoe leaves as specific treatment for epilepsy and argued that mistletoe from any deciduous tree would exert the same effect as mistletoe from oak [12].

While folk medicine at the end of the 19th century still regarded mistletoe as an indispensable part of the medicine chest [13], academic medicine in the rising scientific age lost more and more interest in mistletoe as a remedy [9].

Oak Mistletoe

Special attention has been given to oak mistletoe since ancient times. Gallic druids regarded the *Viscum* as an 'all-healing' plant when they found it growing on oak trees [3]. Specific medieval manuscripts document the outstanding role of oak mistletoe in later periods of European medical history [14, 15], when it was still deemed to have stronger effects than mistletoe from other host trees [11].

According to Pliny, the Gallic druids only rarely found mistletoe on oaks [3]. Several authors [16, 17] even doubted that *V. album* grew on European oaks (*Quercus robur* and *Q. petraea*). They assumed that Pliny, and later medieval authors, might instead have been referring to *Loranthus europaeus*, a parasitic plant that frequently grows on oaks. However, they missed the fact that *Loranthus* was geographically confined to south-eastern Europe [1] and thus was not available in Gaul.

In the 19th century, pharmaceutical botanists became convinced that druids in Gaul used *V. album* that they harvested from *Quercus* trees [18]. Actually, Tubeuf confirmed about 50 oaks in France as hosts of *V. album* after surveying the occurrence of mistletoe in Europe at the beginning of the 20th century [1]. Later, extensive searching by a Swiss research group identified more than 200 *Viscum*-bearing European oaks in France [19]. The existence of several mistletoe-bearing oaks has also been confirmed in England, from where druidism radiated to the European continent [20]. *Viscum* on *Quercus* was documented as well, although in smaller amounts, in Germany and Switzerland [1, 21]. Assuming that the white-berried mistletoe may have been more abundant on European oaks in medieval and ancient times than today, it is safe to imagine that the Gallic druids found a sufficient number of sacred oak trees to satisfy their demand for an *omnia sanans*. However, *V. album* from oak might also have been available to medical specialists in other parts of Europe.

To avoid confusion, it is important to consider the distinct morphological differences between the two types of oak mistletoe. In the 18th century, a German encyclopaedia emphasised that the oak mistletoe, which was used for medicinal purposes, is green both in summer and winter, has white berries and shows elongated leaves with parallel veins [11]. These features are characteristic of *V. album*. In contrast,

Loranthus europaeus is leafless in winter, has bright yellow berries and shows more round leaves. Further characterisation of the oak mistletoe was recommended by noting the wood to which it is attached [11], taking into account that *Viscum* grows on a wide range of deciduous host trees like maple, poplar, birch, lime, apple, elm, and ash as well as on coniferous trees like pine and fir [22]. *Loranthus*, on the other hand, grows almost exclusively on oaks and can only be found in southern parts of Europe. While druids in Gaul surely used *V. album*, ancient and medieval doctors in other parts of Europe might have applied *L. europaeus* as a substitute when *V. album* on oak was not available in their region [23].

Introduction of Mistletoe into Cancer Therapy

By the end of the 19th century, new interest in mistletoe arose, illustrated by its wide use as an artistic motif in the *Jugendstil* or *Art Nouveau* movements [24]. At the beginning of the 20th century, Tubeuf [1] collected knowledge about *V. album* from past millennia in a scientific monograph and combined this knowledge with more recent data about its abundance in Europe. At the same time, medical interest in mistletoe was revived by Gaultier, who used aqueous extracts of *V. album* for pills and subcutaneous injection against high blood pressure [25, 26].

In 1917, Ita Wegman, one of the first women to study and practise medicine in Switzerland, began to treat cancer patients in the Woman's Hospital of Zurich with mistletoe extracts after she had elaborated – in close collaboration with an experienced pharmacist – an injectable solution from dried mistletoe plants [27]. Until then, mistletoe had not been specifically used for tumour treatment [28]. One has to also consider that chemotherapy was not available at that time because the first cytostatic drugs were developed after World War II. Surgery was comparatively crude, and radiotherapy was not only rather harmful, but still in an experimental stage.

The underlying concept to use mistletoe in cancer therapy was developed by Rudolf Steiner, the founder of anthroposophy. Based on Wegman's practical experience after 3 years of clinical observations, he described *V. album* as a specific remedy for cancer in the spring of 1920, when he introduced the basic elements of anthroposophical medicine to medical doctors and homoeopaths [29]. Steiner went on to develop the initial concept further, proposing specific harvesting periods for mistletoe in summer and winter plus an ambitious technical process to combine the respective extracts. He also suggested the development of differentiated mistletoe therapies for different types of cancer that were based on the variety of host trees from which *V. album* was gathered.

The introduction of mistletoe as a remedy for cancer differs from the way other plant ingredients were introduced into modern cancer therapy. Paclitaxel (Taxol), a plant substance from the Pacific yew tree *(Taxus brevifolia)* that became important in breast cancer therapy at the end of the 20th century, was discovered in 1962 in a screening programme of more than 200 different plant species, which was supported by the U.S. National Cancer Institute [30]. Mustard oil, a substance that is found in cruciferous plants and is especially abundant in broccoli, has been regarded as beneficial since the late 1970s, when epidemiological studies found a possible link between diets high in cruciferous vegetables and a lower risk of cancer, a theory that subsequently was supported by animal experiments [31].

Thinking holistically, Steiner understood, on the one hand, that tumour growth was based on a loss of balance within the human organism [29] and that external or internal influences could enhance the imbalance between vital and formative forces, thus creating an environment of uncontrolled cell proliferation and tumour growth. On the other hand, he meticulously studied the qualities of mistletoe and came to the conclusion that it could counter the hypertrophy of cancer cells. He postulated that specifically processed mistletoe extracts could re-establish the body's own ability to regain control and thereby help to create a new balance within the human organism [32].

Methodological Roots

Steiner's methodological approach can be traced back to 1879, when he enrolled for studies of natural science at Vienna's Technical University [33]. Scientific botany at that time was dealing with the question of why roots showed gravitropism and clearly distinguished mistletoe's haustorium from the normal root [34]. Aware of subsequent scientific progress, Steiner later pointed to the fact that *V. album* has no organ that could function to respond to gravity like the statocytes in the root cap of normal higher plants [35]. When introducing his mistletoe-cancer concept and describing specific morphological patterns of *V. album* as being opposite to normal plant growth patterns, he also emphasised the toxicity of mistletoe and declared this quality as essential for its anti-cancer potential [29, 36].

Another root of Steiner's concept is probably closely related to his friendship with an old herbalist, who regularly delivered medicinal herbs to pharmacies in Vienna and whom he had met during his first academic year [37]. On long walks in nature, this man introduced Steiner to traditional plant wisdom and helped him to become aware of the inner qualities of plants with medical applications. The herbalist was familiar with all kinds of traditional knowledge and might also have talked to him about specific druidic practices that were preserved by oral tradition in Styria. According to Ferk, who, as a linguist, analysed the already-mentioned polar aspects of druidism in the Austrian Alps, the black-dressed druid represented winter, while the white-dressed druid represented summer [6]. This mythological representation of polar seasonal qualities may have inspired Steiner when he designed a pharmaceutical process based on harvesting mistletoe in summer and winter [33].

The methods of Steiner's mistletoe-cancer-concept originated in the idea of metamorphosis that Goethe, after several years of non-biased plant observation, had enunciated in 1790. Having edited Goethe's natural scientific writings in the 1880s [38], Steiner further developed the concept of metamorphosis and later formed this idea into practical applications for different areas of daily life like agriculture, pedagogy and medicine [39]. In contrast to Linné and his botanical systematics, Goethe was searching for the archetype of the plant, the so-called *Ur-Pflanze* [40]. He identified the three stages of development in higher plants – vegetative, flowering and fruiting – and saw this as a differentiated process of threefold expansion and contraction. Of specific interest is Goethe's detailed description of the leaf transformation, beginning from small cotyledons, followed by stepwise expansion and differentiation into typically shaped leaves, before final contraction again as rather simple sepals. Compared with this normal shoot development, the shoot growth of *V. album* is strongly reduced: in a year's time from each axillary bud, only one stem unfolds with a terminal pair of simple and undifferentiated leaves [19].

Goethe summarised his observations on the concept of polarity and *steigerung* (intensification). He understood that polar forces of expansion and contraction interacted, for example, in the vegetative growth of the root and green shoot, and intensified in the generative development of the flower, fruit and seed. Steiner applied this concept, integrating traditional wisdom with modern science, when he created a new pharmaceutical concept for processing mistletoe [33]: Extracts of summer and winter mistletoe, which have different pharmacological properties [41; see Urech and Baumgartner, this book], are combined in a high-speed mixing process, which facilitates the formation of new biological qualities [42].

Steiner's concept of using mistletoe from different host trees for the treatment of different types of cancer [29, 43] is related to scientific observations that the haustorium of *V. album* is connected to the xylem of the host tree [44, 45] and receives a host-specific composition of mineral and organic substances [46]. In the early years of cancer-specific mistletoe therapy, *V. album* from *Quercus* was considered as a highly valuable remedy. Soon after Tubeuf [1] had published the site information about mistletoe-bearing oaks, Wegman announced, in 1927, that oak mistletoe from France was available as a remedy for cancer treatment [47].

Later, a Swiss research group related to the Society for Cancer Research launched a long-term programme to safeguard the supply of oak mistletoe [19]. In addition to searching out and cataloguing wild sites in France where the white-berried mistletoe grew naturally on oaks (fig. 1), attempts were initiated to cultivate *V. album* on *Q. robur* and *Q. petraea*. Careful observation of oak disposition for mistletoe and sustainable efforts to establish mistletoe seedlings on oaks resulted in successful methods of oak mistletoe propagation. In the meantime, host-tree husbandry generated more than a hundred young European oaks from which the white-berried mistletoe could regularly be harvested. The results of new clinical studies, in which patients with pan-

Fig. 1. Mistletoe-bearing oak
(Quercus robur) in France
(Photo: K. Urech).

creatic cancer [48, 49; see Galun and Tröger, this book] or patients suffering from cancer fatigue [50; see Zänker, this book] were treated with oak mistletoe extracts, indicated the potential value of these efforts.

Conclusion

One hundred years after being introduced into cancer therapy, the understanding and use of mistletoe continues to grow. This unique plant that, for several thousand years, had been used as a remedy for several different diseases, has now proven its specific potential, especially for integrative cancer treatment and as an element of supportive care. The results of far more than 100 clinical studies have indicated that mistletoe has several beneficial effects in cancer treatment [51]. In complementary medicine in Europe, mistletoe extracts are among the remedies most often used for cancer treatment.

References

1 Tubeuf K: Monographie der Mistel. Berlin, München, Oldenbourg, 1923.

2 Jacomet S, Brombacher C: Abfälle und Kuhfladen – Leben im neolithischen Dorf. Zu Forschungsergebnissen, Methoden und zukünftigen Forschungsstrategien archäobotanischer Untersuchungen von neolithischen Seeufer- und Moorsiedlungen. Jahrbuch der Schweizerischen Gesellschaft für Ur- und Frühgeschichte 2005;88:7–39.

3 Bayer G, Brodersen K, Hopp J, König R, Winkler G (eds): Plinius C. Secundus d. Ä., Naturkunde, Buch 16:Botanik: Waldbäume. München, Zürich, Artemis und Winkler, 1994.

4 Virgil: The Aeneid. Project Gutenberg, 2008. http://www.gutenberg.org/files/228/228-h/228-h.htm#book06 (accessed March 9, 2015).

5 Ramm H, Buess J: Zaubermistel – goldener Zweig. Basel, Futurum, 2013.

6 Ferk F: Über Druidismus in Noricum. Graz, Leuschner & Lubensky, 1877.

7 Snorre: The Younger Edda. Project Gutenberg, 2006, chapt XV – The Death of Balder. http://www.gutenberg.org/files/18947/18947-h/18947-h.htm (accessed March 9, 2015).

8 Celsus AC: De Medicina, transl by Spencer WG. London, William Heinemann Ltd., Cambridge, Massachusetts, Harvard University Press, 1961, vol II. https://ia700401.us.archive.org/23/items/demedicina02celsuoft/demedicina02celsuoft_bw.pdf (accessed March 9, 2015).

9 Büssing A: History of mistletoe uses; in Büssing A (ed): Mistletoe. The Genus Viscum. Amsterdam, Harwood Academic Publishers, 2000, pp 1–6.

10 Dormandy T: The Worst of Evils: The Fight against Pain. New Haven, CT, Yale University Press, 2006.

11 Grosses Universallexikon der Wissenschaften und Künste. Halle, Leipzig, Zedler, 1739, vol 21 (Mi-Mt). http://www.zedler-lexikon.de/index.html?c=blaettern&id=189592&bandnummer=21&seitenzahl=0274&supplement=0&dateiformat=1' (accessed March 9, 2015).

12 Colbatch J: A Dissertation Concerning Mistletoe – A Most Wonderful Specific Remedy for the Cure of Convulsive Distempers, ed 4. London, Dan. Browne, 1725.

13 Kneipp S: Meine Wasserkur. Kempten, Verlag d. Josef Kösel'schen Buchhandlung, 1891.

14 Telle J: Altdeutsche Eichentraktate. Centaurus 1968;13:37–61.

15 Högemann A, Keil G: Der 'Straßburger Eichentraktat' – ein zum Wunderdrogen-Text gewordenes Albertus-Magnus-Kapitel; in Engelhardt H, Kempter G (eds): Diversarium artium studia: Beitr. zu Kunstwiss., Kunsttechnologie u. ihren Randgebieten. Wiesbaden, Reichert, 1982, pp 267–276.

16 Paul H (ed): Wörterbuch der deutschen Pflanzennamen von Heinrich Marzell. Stuttgart, Hirzel, 1979.

17 Genaust H: Etymologisches Wörterbuch der botanischen Pflanzennamen, ed 3. Basel, Birkhäuser, 1996.

18 Geiger PL: Pharmaceutische Botanik, ed 2. Heidelberg, Winter, 1839.

19 Ramm H, Urech K, Scheibler M, Grazi G: Cultivation and development of Viscum album L.; in Büssing A (ed): Mistletoe, The Genus Viscum. Amsterdam, Harwood Academic Publishers, 2000, pp 75–94.

20 Box JD: Mistletoe Viscum album L. (Loranthaceae) on oaks in Britain. Watsonia 2000;23:237–256.

21 Perrot J: Immortelle verdeur. La Salamandre 2003;157:25.

22 Barney CW, Haksworth FG, Geils BW: Hosts of Viscum album. Eur J For Path 1998;28:187–208.

23 Birkhan H: Kelten. Bilder ihrer Kultur. Wien, Verlag der Österreichischen Akademie der Wissenschaften, 1999.

24 Becker H, Schmoll gen. Eisenwerth H: Mistel – Arzneipflanze, Brauchtum, Kunstmotiv im Jugendstil. Stuttgart, Wissenschaftliche Verlagsgesellschaft mbH, 1986.

25 Gaultier R: Résultats cliniques et expérimentaux de quelques études sur la valeur thérapeutique et physiologique du gui de chêne. Bull Gén Thérap Méd et Chir (Paris) 1906;152:67–72, 88–111, 141–146.

26 Gaultier R: Études nouvelles sur le gui considéré comme médicament hypotenseur. Bulletin Général de Thérapeutique 1912;63:593–620, 641–664.

27 Daems FW: Ita Wegman und das erste Mistelpräparat Iscar zur Krebsbehandlung; in Leroi R (ed): Misteltherapie. Eine Antwort auf die Herausforderung Krebs. Die Pioniertat Rudolf Steiners und Ita Wegmans. Stuttgart, Freies Geistesleben, 1987, pp 35–44.

28 Bellmann PG, Daems FW: Ist die Mistel ein altes Krebsheilmittel? Sudhoffs Archiv für Geschichte der Medizin und der Naturwissenschaften 1965;49:355–363.

29 Steiner R: Geisteswissenschaft und Medizin, GA 312, ed 7, 13th lecture, Dornach, 2. April 1920. Dornach, Rudolf Steiner Verlag, 1999.

30 Paclitaxel. http://en.wikipedia.org/wiki/Paclitaxel (accessed September 23, 2014).

31 Mustard Oil. http://en.wikipedia.org/wiki/Mustard_oil (accessed September 23, 2014).

32 Steiner R: Anthroposophische Menschenerkenntnis und Medizin, GA 319, ed 3, 3rd lecture, London, 3. September 1923. Dornach, Rudolf Steiner Verlag, 1994.

33 Ramm H: Rudolf Steiner und das Wesen der Mistel. Mistilteinn – Beiträge zur Mistelforschung 2011;9:4–25.

34 Sachs J: Vorlesungen über Pflanzenphysiologie. Leipzig, Wilhelm Engelmann, 1882.

35 Steiner R: Antworten der Geisteswissenschaft auf die grossen Fragen des Daseins – Der Geist im Pflanzenreich, GA 60, ed 2, 6th lecture, Berlin, 8. Dezember 1910. Dornach, Rudolf Steiner Verlag, 1983.

36 Steiner R: Zeitgeschichtliche Betrachtungen, GA 173b, special ed, 13th lecture, Dornach, 31. Dezember 1916. Dornach, Rudolf Steiner Verlag, 2011.

37 Selg P: Rudolf Steiner und Felix Koguzki. Arlesheim, Ita Wegman Institut, 2009.

38 Steiner R: Die Entstehung der Metamorphosenlehre; in Steiner R (ed): Goethes Naturwissenschaftliche Schriften. Stuttgart, Berlin, Leipzig, Union Deutsche Verlagsgesellschaft, 1883, vol I.

39 Kries M, Vegesack A (eds): Rudolf Steiner: Alchemy of the Everyday. Weil a. Rhein, Vitra Design Museum, 2010.

40 Larson JL: Goethe and Linnæus. Journal of the History of Ideas 1967;28:590–596.

41 Urech K, Jäggy C, Schaller G: Spatial and seasonal variations of viscotoxins and mistletoe lectins in mistletoe (*Viscum album* ssp., *album*) and their consideration in anthroposophical pharmacy (Abstract). Forschende Komplementärmedizin 2007;14(suppl 1): 30.

42 Baumgartner S, Flückiger H, Kunz M, Scherr C, Urech K: Evaluation of preclinical assays to investigate an anthroposophic pharmaceutical process applied to mistletoe (*Viscum album* L.) extracts. Evid Based Complement Alternat Med 2014;2014: 620974.

43 Walter H: Der Krebs und seine Behandlung – Eine Sammlung von Krankengeschichten mit Hinweisen von Dr. Rudolf Steiner. Stuttgart, Arlesheim, Verein für Krebsforschung, 1953.

44 Sallé G: Germination and establishment of *Viscum album* L.; in Calder DM, Bernhardt P (eds): The Biology of Mistletoes. London, Academic Press, 1983, pp 145–159.

45 Becker H: Botany of European mistletoe (*Viscum album* L.). Oncology 1986;43(suppl)1:2–7.

46 Ramm H: Einfluss bodenchemischer Standortfaktoren auf Wachstum und pharmazeutische Qualität von Eichenmisteln (*Viscum album* auf *Quercus robur* und *petraea*). Landbauforschung Völkenrode FAL Agricult Res, Sonderheft 301, 2006.

47 Urech K: Die Eiche von Isigny-le-Buat – Wahrzeichen der Eichenmistel in Frankreich. Mistilteinn 2002;3:4–13.

48 Galun D, Tröger W, Reif M, Schumann A, Stankovic N, Milicevic M: Phase III trial on mistletoe extract versus no antineoplastic therapy in patients with locally advanced or metastatic pancreatic cancer. Eur J Integrat Med 2012;4(suppl):11–12.

49 Tröger W, Galun D, Reif M, Schumann A, Stankovic N, Milicevic M: *Viscum album* [L.] extract therapy in patients with locally advanced or metastatic pancreatic cancer: A randomised clinical trial on overall survival. Eur J Cancer 2013;49:3788–3797.

50 Bock PR, Hanisch J, Matthes H, Zänker KS: Targeting inflammation in cancer-related fatigue: a rationale for mistletoe therapy as supportive care in colorectal cancer patients. Inflamm Allergy Drug Targets 2014;13:105–111.

51 Kienle GS, Kiene H: Die Mistel in der Onkologie. Stuttgart, Schattauer, 2003.

Hartmut Ramm
Hiscia Institute, Society for Cancer Research
Kirschweg 9
CH-4144 Arlesheim (Switzerland)
E-Mail h.ramm@vfk.ch

Zänker KS, Kaveri SV (eds): Mistletoe: From Mythology to Evidence-Based Medicine.
Transl Res Biomed. Basel, Karger, 2015, vol 4, pp 11–23 (DOI: 10.1159/000375422)

Chemical Constituents of *Viscum album* L.: Implications for the Pharmaceutical Preparation of Mistletoe

Konrad Urech · Stephan Baumgartner

Hiscia Institute, Society for Cancer Research, Arlesheim, Switzerland

Abstract

Early results of the pharmacological effects of *Viscum album* have challenged the field of phytochemistry to identify active compounds. A broad spectrum of different proteins, polysaccharides, phenolics, lipophilic molecules and other compounds has been detected in the three European subspecies of *V. album*. The results of the chemical constituents of *V. album* have been compiled in this review, which illustrates the chemical and pharmacological potential of this plant. The spatiotemporal distribution of the main active compounds, the viscotoxins and mistletoe lectins, which are responsible for the characteristic and chemical features of *V. album*, is shown. The impact of these chemical results on pharmaceutical processing and quality control of mistletoe extracts is described in detail.

© 2015 S. Karger AG, Basel

Introduction

Proprietary mistletoe (*Viscum album* L.) preparations for use in cancer therapy are composed of total plant extracts. Since 1917, when the administration of total mistletoe extracts as therapy options in cancer patients began, positive 'healing' effects have been observed [1]. These early and on-going clinical observations stimulated the analytical research of *V. album* [2]. Numerous preclinical studies have identified an increasing number of compounds in *V. album* that are supposed to play roles as active compounds in cancer therapy and prevention. This review gives an overview of the compounds and metabolites of the different chemical classes found in European *V. album*, with reference to plant-derived chemicals with antitumour potential. The results of the spatiotemporal distribution of the main active ingredients of *V. album* demonstrate the spatial and seasonal dynamics of the metabolic activity of this species. The significance of the accumulated chemical data of European *V. album* for pharmaceutical processing is discussed.

Chemical Diversity and Features of *Viscum album*

Phytochemistry and Pharmacology of Mistletoe: Historical Aspects
Pronounced pharmacological effects of mistletoe extracts were discovered and published in the scientific literature in the 19th century and at the beginning of the 20th century. Hitherto, the unknown lethal effects in animals [3] and observational studies in patients on the recovery of granulating wounds [4], cardiovascular effects [5] and anticancer effects [6] of mistletoe illustrate a wide range of pharmacological phenomena of mistletoe in intact organisms and has stimulated experimental work on active ingredients of mistletoe.

Cardiovascular effects were the first to be associated with chemical compounds in mistletoe. Choline, acetylcholine [7] and γ-aminobutyric acid [8] were identified as key molecules for a transient hypotensive effect. Around 1950, a new class of mistletoe proteins was described, named viscotoxins, which could induce hypotension by cardiovascular toxicity [9].

Mistletoe is a food for domestic animals such as cattle and wild animals, but parenteral feeding is lethal [3]. Koch [10] observed two principles of lethal toxicity: mice died either immediately after injection or 2–4 days later. This observation was a first clue of the presence of two different groups of toxic compounds in mistletoe: the viscotoxins and the mistletoe lectins (MLs).

The first and most remarkable approach to isolate mistletoe compounds with antitumour activity was based on a mouse model of implanted Crocker sarcoma 180. Lipoid, polysaccharide and protein fractions of mistletoe were identified as active constituents [11]. A protein complex composed of about 10 proteins of different molecular weights ('Vester's proteins', VP16 [12]) was identified and represented the first antitumoural proteins known at that time. Further fractionation of this complex resulted in loss of antitumour activity. Retrospective analysis of these data showed that MLs were the main component of VP16 [13]. In the 1950s, observed haemagglutination gave a first hint to the presence of lectins in *V. album*. However, 15 years later, analysis of the sugar-binding proteins was initiated. The consecutively detected antiproliferative, immunomodulating and antitumour properties characterised the MLs as the most important and best-known active ingredients of *V. album*.

Koch [10] developed a new anticancer treatment on the basis of the necrotic effects of mistletoe extracts using murine systems. He demonstrated the antigenic nature of the active compounds, but, at that time, he could not identify the molecules. Later on, necrogenesis was linked to the viscotoxins. The molecular structures and the antiproliferative effects of the viscotoxins are now state-of-the-art, current knowledge. Part of the antitumoural activity of mistletoe is attributed to its polysaccharide fractions [11]. Therefore, Müller [14] was able to produce mistletoe-derived polysaccharides showing antitumoural activity in murine models. However, the polysaccharide fractions were only partially chemically characterised. Moreover, these preparations stimulated human bone marrow by inducing neutrophilia in otherwise neutropenic patients [15].

'Mistletoe resin' named 'viscin', which was used for centuries as 'bird lime', attracted the interest of scientists in the 19th century [16]. Traditional therapeutic applications of this lipophilic and sticky substance had been described in the ancient world, and beneficial effects of a viscin ointment on atopic eczema, burns and granulating wounds were reported in 1906 [4]. Later, the presence of oleanolic acid and other triterpenes were discovered [17, 18]. The antitumoural potential and a wide range of other pharmacological activities of triterpenes are now well established. A case series study on the topical application of a total lipophilic extract of mistletoe has recently been shown to be effective in the treatment of basal cell carcinoma [19].

Chemical Compounds in Viscum album
The chemical compounds identified in *V. album* have been compiled with the corresponding classification and references in table 1.

Viscotoxins
Viscotoxins show the structural characteristics of plant α- and β-thionins, which are cysteine-rich and highly basic proteins of about 5 kD in molecular mass. Seven different isoforms have been isolated and identified in *V. album*: viscotoxin A1, A2, A3, B, B2, C1 and 1-PS, each consisting of 46 amino acids with high sequence homology. Thirty-two of the 46 positions have identical amino acids, and all isoforms contain three disulphide bridges at highly conserved positions (Cys3/Cys40, Cys3/Cys32, and Cys16/Cys26), giving them a compact structure. This probably explains their high stability under denaturing conditions (heat and proteases). Analysis of the 3D-structures of these proteins also yielded information on a specific phosphate-binding site [20] that was discovered by Orrù et al. [21]. This phosphate binding site as well as amphipathic structures of the viscotoxins might interfere with cell membranes and destroy their integrity, leading to the cytotoxic effects of the viscotoxins, primarily necrosis and less so induction of apoptosis [22]. Interestingly, in spite of their high structural identity, the different viscotoxin isoforms show different biological behaviours and differential distributions in the 3 subspecies of *V. album* [23–25].

The viscotoxins show a high structural and pharmacological relationship to snake (cobra) cardiotoxins [26]. However, U-PS, formerly thought to belong to the viscotoxin family, turned out to be a γ-thionin [57].

Table 1. Chemical compounds identified in the European *Viscum album* L.

Class of chemical compounds		Compounds	References
Proteins	Thionins	Viscotoxins A1, A2, A3, B, B2, C1, 1-PS	[53–57]
	Lectins	Mistletoe lectin I (MLI), II (MLII), III (MLIII) Chitin-binding mistletoe lectin 1 (cbML1), 2 (cbML2), 3 (cbML3)	[58, 59] [29]
Peptides	Oligopeptides	Glutathione	[60]
Amino acids		Arginine, cystein, γ-aminobutyric acid	[8, 61, 62]
Amines		Acetylcholine, choline, tyramine, histamine	[63, 64]
Proteoglycans		Arabinogalactan-proteins	[40]
Polysaccharides		Methylated poly-1→α4 galacturonic acid, arabinogalactan, rhamnogalacturonan, xyloglucan	[31, 65, 66]
Fatty acids		Oleic acid, palmitic acid, linoleic acid, linolenic acid, arachidic acid, cerotic acid, stearic acid, lignoceric acid	[18, 64, 67]
Phenolic compounds	Flavonoids	2′-hydroxy-4′,6′-dimethoxychalcone-glucosid, 2′-hydroxy-4′,6′-trimethoxychalcone-glucosid, 2′-hydroxy-4′,6′-dimethoxy-chalcone-4-O-[apiosyl(1→2)] glucoside, (2R)-5,7-dimethoxyflavanone-4′-O-glucosid, (2S)-3′,5,7-trimethoxy-flavanone-4′-O-glucoside, homoeriodictyol-7-O-glucoside, rhamnazin-3,4′-di-O-glucoside, 5,7-dimethoxy-4′-hydroxyflavon *After acid hydrolysis*: homoeriodyctiol, sakuranetin, rhamnazin, isorhamnetin, quercetin, 6 different quercetin methylesters, kaempferol, naringenin	[36, 68–70]
	Phenylpropanoids	Syringenin-4′-O-glucoside (syringin), syringenin-4′-O-apiosyl-1→2 glucoside (syringoid), syringaresinol-4,4′-O-glucoside, eleutheroside E, syringaresinol-mono-O-glucoside, sinapic acid, cinnamic acid, rosmarinic acid, caffeic acid, ferulic acid, chlorogenic acid, isochlorogenic acid, syringic acid, p- and m-coumaric acid	[68, 69, 71]
	Other phenolic acids	Gallic acid (3,4,5-trihydroxybenzoic acid), digallic acid, para-OH benzoic acid, syringic acid (methylated trihydroxybenzoic acid), protocatechuic acid (3,4-dihydroxybenzoic acid), vanillic acid (methylated dihydroxybenzoic acid), gentisic acid (2,5-dihydroxybenzoic acid), salicylic acid (2-hydroxybenzoic acid), ellagic acid	[69, 72]
Terpenoids	Triterpenoids	Oleanolic acid, betulinic acid, ursolic acid, β-amyrin, β-amyrin acetate, lupeol, lupeol acetate	[18, 68]
	Tetraterpenoids	Carotin	[73]
Phytosterols		β-sitosterol, stigmasterol	[18]
Inorganic substances		Manganese, potassium, calcium (calcium oxalate)	[38, 39]
Various compounds		Ascorbic acid 7-iso-jasmonic acid, and its precursor 12-oxophytodienoic acid	[41] [42]

Pharmacologically active and characteristic mistletoe compounds and the corresponding references have been selected and tabulated according to their chemical classification.

Lectins

Mistletoe Lectins. Three different MLs with differential sugar-binding specificities were isolated and identified in *V. album*: galactose-specific MLI (115 kDa, dimer), galactose- and N-acetyl-D-galactosamine-specific MLII (60 kDa) and *N*-acetyl-D-galactosamine-specific MLIII (60 kDa). These MLs belong to the type 2 ribosome-inactivating proteins that consist of a lectin subunit B and a toxophoric A-chain, which is an RNA *N*-glycosidase. MLs block protein synthesis by hydrolysing the 28S rRNA in ribosomes of eukaryotic cells and inducing apoptotic cell death. The amino acid sequences, tertiary and quaternary structures of MLI and the A-chain of MLIII have been elucidated [27, 28].

In studies on the 3D-structures of the B chains, the subdomains of MLI and MLIII, which are responsible for sugar binding, were identified, and the N-glycosylation sites were localised. Isoelectric focusing of the three MLs separated more than 20 different isoforms of the MLs in total; these MLs differed mainly in their glycosylation patterns.

Chitin-Binding Mistletoe Lectins. Chitin-binding mistletoe lectins (cbMLs), with a MW of 10.8 kD, were first described in 1996 [29]. The three isoforms, cbML1, cbML2 and cbML3, which have very closely related primary structures, share high structure similarity with hevein.

Polysaccharides

Structure elucidations of mistletoe polysaccharides showed differences between the fruit and green parts (leaves and stems) of mistletoe. A highly methylated galacturonan, a pectin with a molecular weight of 42 kD (after hydrolysis of ester bonds), and arabinogalactan, with a molecular weight of 110 kD, were isolated from stems and leaves [30]. The berries, which are especially rich in polysaccharides, contain rhamnogalacturonans with arabinogalactan side chains (up to 1,340 kD), arabinogalactans and small amounts of xyloglucans [31]. The high-molecular-weight arabinogalactan selectively stimulates the proliferation of CD4+ T helper lymphocytes [32], and stimulation of natural killer cell activity by Iscador® depends on rhamnogalacturonan [33].

Liposoluble Compounds

V. album is rich in triterpenes [34]: the oleananes oleanolic acid, β-amyrin and β-amyrinacetate; the lupanes lupeol, lupeol acetate and betulinic acid; and the ursane ursolic acid. The phytosterols stigmasterol and β-sitosterol are also present in *V. album* [18]. Lipophilic extracts contain the saturated fatty acids palmitic, arachidic, lignoceric, behenic and cerotic acids and the unsaturated oleic, linoleic and linolenic acids [18, 35].

Flavonoids, Phenylpropanoids, and Phenolic Acids

The phenolic compounds flavonoids, phenylpropanoids and phenolic acids share common features in their biosynthetic pathways. They represent one of the most abundant classes of substances in the plant kingdom and are present in *V. album* (table 1). Evidence of their pharmacological roles in a wide variety of pathologies is of increasing interest (H. Schramm, Chapter III). The three European subspecies of *V. album* differ in flavonoid composition [36]. When testing mistletoe preparations, it was suggested that syringenin-apiosylglucoside should be used for fingerprint identity [37]. The well-known flavonoid quercetin, in its methylated form, could be identified in *V. album* only after acid hydrolysis.

Inorganic Substances
Mistletoe is comprised of high amounts of mineral ions. Potassium is enriched from the host by a factor of about 10. Calcium is mainly present in its non-soluble oxalate form [38]. Manganese can be found in higher concentrations only in *V. album* growing on oak and fir. Host specificity reflects the selective uptake of soil minerals by the different host species [39].

Other Bioactive Compounds
Arabinogalactan Proteins. The recently isolated arabinogalactan proteins have arabinose:galactose ratios of 1:0.7 (berries) and 1:1.18 (herb) [40]. However, the protein moiety has not been characterised. Cross-reaction with Echinacea arabinogalactan protein antibodies demonstrated the existence of cognate mimicry epitopes. This new mistletoe substance also stimulates the toll-like receptor [40].

Ascorbic Acid. High concentrations of 7.5-mg/g ascorbic acid have been detected in mistletoe fruits [41].

Jasmonic Acid. Estival accumulation of the antiproliferative jasmonic acid and its precursor 12-oxophytodienoic acid occurs in *V. album* [42].

Cysteine, Glutathione. The sulphur content represented by cysteine, glutathione and the viscotoxins is especially high in *V. album* compared to its hosts. Furthermore, the concentration of the thiols cysteine and glutathione are low in summer but accumulate in autumn and winter [43]. Together with ascorbic acid and the phenolic compounds, cysteine and glutathione contribute to the antioxidant potential of mistletoe.

Spatiotemporal Variations of Antitumour Compounds in Viscum album
The host species plays an important role in the chemical composition of mistletoe, and two different determining factors have to be taken into account: (a) the uptake of host-specific chemical compositions of the nutrients from the xylem sap and (b) the genetic imprinting by the subspecies of *V. album*. *V. album* ssp. *album* is strictly bound to deciduous trees, *V. album* ssp. *abietis* Beck is bound to hosts of the genus *Abies,* and *V. album* ssp. *austriacum* (Wiesb.) Vollmann is bound to *Pinus* sp., but is rarely bound to *Picea abies* [44]. Host-specific quantitative and qualitative differences in the patterns of the antitumour viscotoxin isoforms in the three European subspecies of *V. album* have been observed in a comparative study [25]. The subspecies could be identified by their viscotoxin patterns not only in fresh

Table 2. Total viscotoxin contents (mg/g fresh weight) and proportions of the viscotoxin isoforms A1, A2, A3, B, U-PS and 1-PS in the three European subspecies of *V. album* on different host species

Viscotoxin isoforms	*V. album* ssp. *album*						*V. album* ssp. *abietis*		*V. album* ssp. *austriacum*	
	Host: *Malus domestica*		Host: *Quercus* sp.		Host: *Ulmus* sp.		Host: *Abies pectinata*		Host: *Pinus sylvestris*	
	Fresh plant (n = 10)	Iscador (n = 6)	Fresh plant (n = 12)	Iscador (n = 6)	Fresh plant (n = 10)	Iscador (n = 6)	Fresh plant (n = 10)	Iscador (n = 6)	Fresh plant (n = 12)	Iscador (n = 6)
A1	15.8±3.5	18.2±1.5	14.1±2.6	19.7±3.1	16.0±3.4	16.6±2.9	6.1±2.3	6.5±0.6	n.d.	n.d.
A2	35.0±3.6	30.0±2.7	34.3±6.4	28.3±2.9	34.0±6.7	40.2±2.4	n.d.	05±1.1	1.2±1.5	1.4±1.3
A3	39.4±3.3	38.3±1.89	41.2±5.0	44.9±3.4	36.1±5.9	37.1±1.6	74.5±4.0	73.4±1.7	6.1±4.8	0.9±0.4
B	9.3±4.2	13.5±2.6	11.0±5.0	7.1±2.8	14.0±7.4	6.1±2.9	5.4±4.2	3.0±1.5	14.2±9.1	13.9±4.2
1-PS	n.d.	n.d.	n.d.	n.d.	n.d.	n.d.	13.8±2.9	16.6±2.2	37.9±6.5	32.8±8.6
U-PS	n.d.	n.d.	n.d.	n.d.	n.d.	n.d.	n.d.	n.d.	40.6±11.0	51.1±10.1
Total viscotoxin, mg/g fw	2.9±0.8	2.1±0.1	4.0±0.5	3.0±1.0	4.2±1.3	3.2±0.1	4.0±1.2	2.4±0.4	1.5±0.3	1.2±0.1

Extracts of *V. album* (leaves and stems) and Iscador® were analysed, and the mean values ± SD per g extracted plant were calculated. N = Number of samples; n.d. = not detectable.

plant extracts but also in their fermented products, (Iscador®) (table 2). The different pharmacologies of the various viscotoxin isoforms [45] mirror the provenance of mistletoe.

The spatial distributions of viscotoxins and lectins were measured by Urech et al. [46] (fig. 1). They found that the leaves contain high concentrations of viscotoxins and low concentrations of MLs. However, with increasing age, the stems accumulate higher concentrations of MLs, whereas the viscotoxin concentration decreases. The generative organs contain high amounts of both viscotoxins and MLs.

The seasonal dynamics of the antitumour proteins in the leaves of *V. album* show a pronounced peak of maximal viscotoxin concentrations in June, whereas the MLs accumulate to a maximum amount in December (fig. 2). June and December, as outstanding moments in the physiological state of mistletoe, correspond to the harvesting seasons for the summer and winter, respectively. Mistletoe extracts processed from both harvesting periods are the most proprietary mistletoe preparations for cancer therapy.

Conclusions and Pharmaceutical Implications

This review focused on the mistletoe compounds with antitumour potential. An impressive number of pharmacologically active metabolites have been identified in *V. album*. These compounds do not belong to a single chemical class but to a broad spectrum of different proteins, polysaccharides, liposoluble compounds, and secondary metabolites.

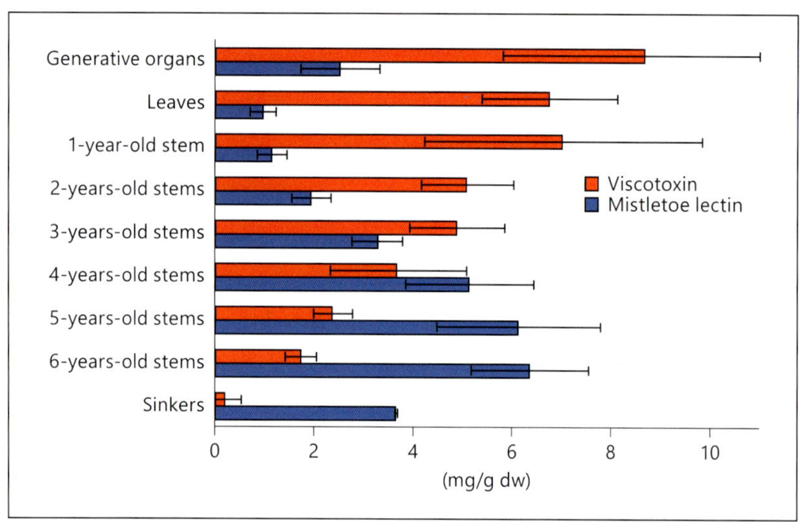

Fig. 1. Distribution of the concentrations (mg/g dry weight) of total viscotoxin and mistletoe lectin in the different green organs of *V. album* growing on *Malus domestica* (mean values of three mistletoe bushes ± SD, adapted from [46]).

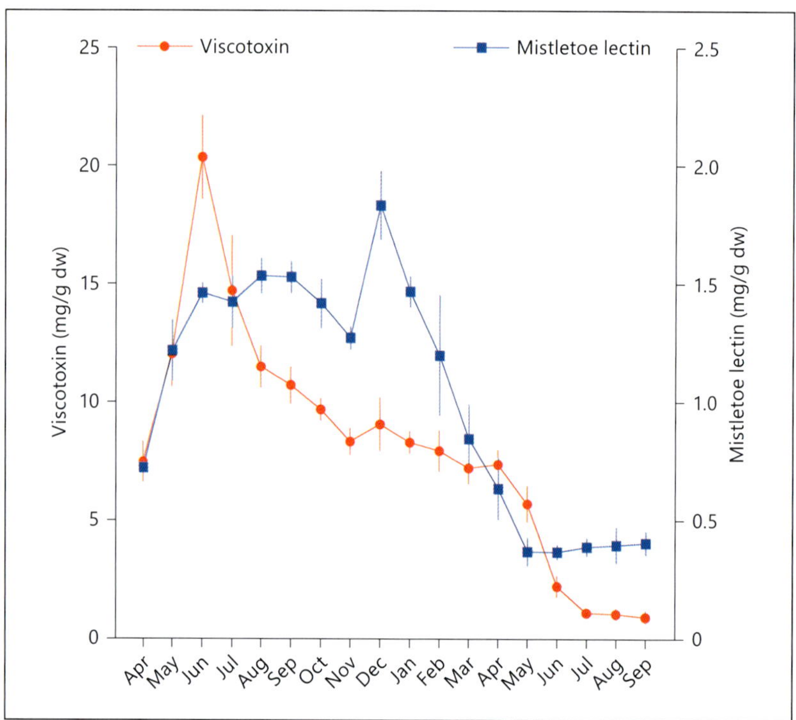

Fig. 2. Seasonal variations of total viscotoxin and mistletoe lectin concentrations in the leaves of *V. album* during the vegetation cycle of leaf development (adapted from [52]).

Special reference has to be given to two protein groups: the viscotoxins and lectins, both of which exert antitumour effects in preclinical systems. As these compounds are unique for *V. album,* they also serve as biochemical markers. Applications of MLs in humans provoked pharmacological phenomena at clinically relevant doses of mistletoe therapy [47]. However, none of the well-known single constituents of *V. album* have been shown to be responsible for the anticancer effects in patients; notably, the hypothesis that the effects of mistletoe therapy were exclusively dependent on MLI could not be corroborated until recently. Obviously, a multitude of potential active substances has to be considered. Due to the lack of a correlation of clinical effects with the application of single active ingredients, it has to be emphasised that, for the time being, the total extract has to be considered 'the active substance'.

The complex and characteristic pattern of chemical compounds is the result of the metabolic activity of the mistletoe plant. This pattern of metabolites gives rise to the broad diversity of pharmacological activities of mistletoe due to single compounds exerting multiple mechanisms of actions on their own; however, they are also interactive molecular partners with other compounds. The interactive potential of various antitumour substances of mistletoe has repeatedly been observed in preclinical assays [e.g. 48–50]. The use of mistletoe total extract, with its enormous complexity of chemical compounds, seems to diverge from the academic mainstream of pharmaceutical approaches using single or, at best, a few defined chemical substances in cancer treatment.

The mistletoe preparations for cancer therapy originate from the view of anthroposophic medicine. The philosophy and holistic view of anthroposophic medicine is described in detail by Kienle et al. [51].

Pharmaceutical processing deeply influences and modifies the spectrum of chemical compounds present in mistletoe extracts as well as the preclinical and clinical effectiveness of these extracts. For example, confinement to aqueous extracting procedures excludes most of the liposoluble compounds. The potential for further development of mistletoe pharmaceuticals by including lipophilic extracts became evident by positive results of an extemporaneous mistletoe product in basal cell carcinoma patients [19].

Mistletoe-specific compounds, as marker substances, may serve as indicators of the reproducible quality of the total extracts. Quality control of the Iscador® products, e.g., includes specifications of viscotoxin concentrations [52]. Total ML content is specified in Plenosol® and Iscador® spec. preparations. Stability assessments of these lead substances are included in quality assurance in order to warrant reproducible pharmaceutical quality.

The complexity and spatiotemporal dynamics of mistletoe metabolites imply the following pharmaceutical requirements for producing total extracts of mistletoe with standardised quality:

- Balancing natural variations by processing large lots of plant material.

- Constant and pharmacologically meaningful proportions of the different green organs of the mistletoe plant.
- Processing a constant proportion of mistletoe fruits (as an important source of polysaccharides).
- Respecting host specificity.
- Respecting seasonal state.
- Blending summer and winter extracts.
- Stabilising active proteins by using total extracts as a matrix but also by avoiding high temperatures.

The modes of clinical application of mistletoe products seem appropriate to their complex pattern of chemical compounds. Mistletoe preparations are usually applied by repeated series of increasing dosages but also by constant, however individualised, determined dosages. These dosage regimens guarantee rhythmic applications of differential optimal doses of the different active compounds in the extracts or allow individual evaluation of the doses with expected effects (e.g. local reactions as a surrogate parameter).

Mistletoe preparations are widely used in therapies for the primary, adjuvant and palliative treatment of cancer.

This review documents the variety and analytics of active antitumour compounds derived from the European mistletoe. However, only the highest analytical and safety standards of the state-of-the-art pharmacognosy and pharmacology fields justify the use of mistletoe (*V. album* L.) in clinical settings.

References

1 Daems WF: Ita Wegman-Zürcher Zeit 1906–1920. Dornach, Verlag am Goetheanum, 1986.

2 Urech K: Die experimentelle Wissenschaft 'entdeckt' die Tumorwirksamkeit der Mistel; in Leroi R (ed): Misteltherapie. Stuttgart, Verlag Freies Geistesleben, 1987, pp 100–118.

3 Gaspard B: Mémoire physiologique sur le gui (*Viscum album*). J Physiol Exp Path 1827;7:227–333.

4 Klug A: Viscolan, eine neue Salbengrundlage. Dtsch Med Wochenschrift 1906;32:2071–2072.

5 Gaultier R, Chevalier J: Action physiologique du gui (*Viscum album*). CR Acad Sc Paris 1907;145:941–942.

6 Wegman I: Die erste Krebsbehandlung mit *Viscum album*; ein Autoreferat von Ita Wegman (1921). Beitr Erw Heilk 1987;40:233–238.

7 Winterfeld K, Kronenthaler A: Zur Chemie des blutdrucksenkenden Bestandteiles der Mistel (*Viscum album*), III. Mitteilung. Arch Pharm 1942;280:103–115.

8 Samuelsson G: Phytochemical and pharmacological studies on *Viscum album* L. III. Isolation of a hypotensive substance: Gamma-Aminobutyric acid. Svensk Farm Tidskr 1959;63:545–553.

9 Winterfeld K, Bijl LH: Viscotoxin, ein neuer Inhaltsstoff der Mistel (*Viscum album* L.). Justus Liebigs Annal Chem 1948;561:107–115.

10 Koch FE: Experimentelle Untersuchungen über entzündung-und nekroseerzeugende Wirkung von *Viscum album*. Z Ges Exp Med 1938;103:740–749.

11 Selawry OS, Vester F, Mai W, Schwartz MR: Zur Kenntnis der Inhaltsstoffe von *Viscum album*, II. Tumorhemmende Inhaltsstoffe. Hoppe-Seyler's Z Physiol Chem 1961;324:262–281.

12 Vester F, Seel A, Stoll M, Müller JM: Zur Kenntnis der Inhaltsstoffe von *Viscum album*. III. Isolierung und Reinigung cancerostatischer Proteinfraktionen. Hoppe-Seyler's Z Physiol Chem 1968;349;125–147.

13 Franz H: Inhaltsstoffe der Mistel (*Viscum album* L.) als potentielle Arzneimittel. Pharmazie 1985;40:97–104.

14 Müller J: Verfahren zur Gewinnung eines Arzneimittels. Dtsch. Patentamt 1962; Auslegeschrift. 1130112, Ciba AG Basel.

15 Mathé G, Schneider M, Amiel J, Cattan A, Schwarzenberg L, Berno M: Stimulation de la neutrophilie par un extrait polysaccharidique de Viscum album. Son utilisation thérapeutique dans les neutropénies. Rev Franç Etudes Clin Biol 1963;8:1017–1020.

16 Auster F, Schäfer J: *Viscum album*. Leipzig, Verlag G. Thieme, 1960.

17 Winterstein A, Hämmerle W: IV. Mitteilung. Über ein Sapogenin aus *Viscum album*. Z Physiol Chem 1931;199:56–64.

18 Urech K, Scher JM, Hostanska K, Becker H: Apoptosis inducing activity of viscin, a lipophilic extract from *Viscum album* L. J Pharmacy Pharmacol 2005; 57:101–109.

19 Kunz C, Heiligtag HR, Hintze A, Urech K: Topische Behandlung des Basalzellkarzinoms mit *Viscum album*, lipophiler Extrakt 10%, Unguentum – Eine Fallserie-Studie; in Scheer R, Alban S, Becker H, Blaschek W, Kemper FH, Kreis W, Matthes H, Schilcher H, Stange R (eds): Die Mistel in der Tumortherapie 3 – Aktueller Stand der Forschung und klinische Anwendung. Essen, KCV Verlag, 2013, pp 315–322.

20 Debreczeni JE, Girmann B, Zeeck A, Krätzner R, Sheldrick GM: Structure of viscotoxin A3: disulfide location from weak SAD data. Acta Cryst 2003;D59: 2125–2132.

21 Orrù S, Scaloni A, Giannattasio M, Urech K, Pucci P, Schaller G: Amino acid sequence, S-S bridges arrangement and distribution in plant tissues of thionins from *Viscum album*. Biol Chem 1997;378:989–996.

22 Büssing A: Biological and pharmacological properties of *Viscum album* L.; in Büssing A (ed): Mistletoe-The Genus VISCUM. Amsterdam, Harwood Academic Publishers, 2000, pp 123–182.

23 Kahle B, Debreczeni JE, Sheldrick GM, Zeeck A: Vergleichende Zytotoxizitätsstudien von Viscotoxin-Isoformen und Röntgenstruktur von Viscotoxin A3 aus Mistelextrakten; in Scheer R, Bauer R, Becker H, Fintelmann V, Kemper F, Schilcher H (Hrsg): Fortschritte in der Misteltherapie. Aktueller Stand der Forschung und klinische Anwendung. Essen, KVC Verlag, 2005, pp 83–97.

24 Urech K, Schaller G, Giannattasio M: Bioassay zur Bestimmung von Viscotoxinen; in Scheer R, Becker H, Berg A (eds): Grundlagen der Misteltherapie. Stuttgart, Hippokrates-Verlag, 1996, pp 111–116.

25 Schaller G, Urech K, Grazi G, Giannattasio M: Viscotoxin composition of the three European subspecies of *Viscum album*. Planta Med 1998;64:677–678.

26 Rosell S, Samuelsson G: Effect of mistletoe viscotoxin and phoratoxin on blood circulation. Toxicon 1966;4:107–110.

27 Krauspenhaar R, Eschenburg S, Perbandt M, Kornilov V, Konareva N, Mikailova I, Stoeva S, Wacker R, Maier T, Singh T, Mikhailov A, Voelter W, Betzel C: Crystal structure of mistletoe lectin I from Viscum album. Biochem Biophys Res Com 1999;257:418–424.

28 Wacker R, Stoeva S, Pfüller K, Pfüller U, Voelter W: Complete structure determination of the A-chain of mistletoe lectin III from *Viscum album* L. ssp. *album*. J Pept Sci 2004;10:138–148.

29 Peumans WJ, Verhaert P, Pfüller U, Van-Damme EJ: Isolation and partial characterization of a small chitin- binding lectin from mistletoe *(Viscum album)*. FEBS Letters 1996;396:261–265.

30 Jordan E, Wagner H: Structure and properties of polysaccharides from *Viscum album* L. Oncology 1986;43(suppl 1):8–15.

31 Edlund U, Hensel A, Frose D, Pfuller U, Scheffler A: Polysaccharides from fresh Viscum album L. Berry extract and their interaction with *Viscum album* Agglutinin I. Arzneim Forsch/Drug Res 2000;50:645–651.

32 Stein GM, Edlund U, Pfüller U, Büssing A, Schietzel M: Influence of polysaccharides from *Viscum album* L. on human lymphocytes, monocytes and granulocytes in vitro. Anticancer Res 1999;19:3907–3914.

33 Müller EA, Anderer FA: Chemical specificity of effector cell/tumor cell bridging by a Viscum album rhamnogalacturonan enhancing cytotoxicity of human NK cells. Immunopharmacology 1990;19:69–77.

34 Jäger S, Trojan H, Kopp T, Laszczyk MN, Scheffler A: Pentacyclic triterpene distribution in various plants – rich sources for a new group of multi-potent plant extracts. Molecules 2009;14:2016–2031.

35 Deliorman D, Orhan I: Fatty acid composition of *Viscum album* subspecies from Turkey. Chem Nat Compounds 2006;42:641–644.

36 Lorch E: Neue Untersuchungen über Flavonoide in *Viscum album* L. ssp. *abietis, album* und *austriacum*. Z Naturforsch 1993;48:105–107.

37 Wagner H, Jordan E, Feil B: Studies on the standardization of mistletoe preparations. Oncology 1986; 43(suppl 1):16–22.

38 Horak O: Vergleichende Mineralstoff-Analysen an einigen Loranthaceae und deren Wirtspflanzen. Z Pflanzenphysiol 1974;73:461–466.

39 Ramm H: Einfluss bodenchemischer Standortfaktoren auf Wachstum und pharmazeutische Qualität von Eichenmisteln (*Viscum album* auf *Quercus robur* und *petraea*). Landbauforschung Völkenrode, Sonderheft 301, 2006.

40 Herbst B, Classen B, Blaschek W: Charakterisierung von Arabinogalaktan-Proteinen aus Viscum album L. Beeren und Kraut; in Scheer R, Alban S, Becker H, Holzgrabe U, Kemper F, Kreis W, Matthes H, Schilcher H (Hrsg): Die Mistel in der Tumortherapie 2. Aktueller Stand der Forschung und klinische Anwendung. Essen, KVC Verlag, 2009, pp 121–132.

41 Rikovski J, Besaric R: Vitamin C content of some indigenous fruits V. Belgrade Soc Chim Bul 1949;14: 129–132.

42 Dorka R, Miersch O, Hause B, Weik P. Wasternack C: Chronobiologische Phänomene und Jasmonatgehalt bei *Viscum album* L.; in Scheer R, Alban S, Becker H, Holzgrabe U, Kemper F, Kreis W, Matthes H, Schilcher H (Hrsg): Die Mistel in der Tumortherapie 2. Aktueller Stand der Forschung und klinische Anwendung. Essen, KVC Verlag, 2009, pp 49–66.

43 Escher P, Eiblmeier M, Hetzger I, Rennenberg H: Seasonal and spatial variation of reduced sulphur compounds in mistletoes (Viscum album) and the xylem sap of its hosts (Populus euramericana and Abies alba). Physiol Plant 2003;117:72–78.

44 Tubeuf C: Monographie der Mistel. München, Berlin, Verlag Oldenbourg, 1923.

45 Schaller G, Urech K, Giannattasio M: Cytotoxicity of different viscotoxins and extracts from the European subspecies of *Viscum album* L. Phytother Res 1996; 10:473–477.

46 Urech K, Jäggy C, Schaller G: Räumliche und zeitliche Dynamik der Viscotoxin- und Mistellektingehalte in der Mistel (*Viscum album* L.); in Scheer R, Alban S, Becker H, Holzgrabe U, Kemper F, Kreis W, Matthes H, Schilcher H (Hrsg): Die Mistel in der Tumortherapie 2 – Aktueller Stand der Forschung und klinische Anwendung. Essen, KVC Verlag, 2009, pp 67–78.

47 Huber R, Rostock M, Goedl R, Lüdtke R, Urech K, Buck S, Klein R: Mistletoe treatment induces GM-CSF- and IL-5 production by PBMC and increases blood granulocyte- and eosinophil counts: a placebo controlled randomized study in healthy subjects. Eur J Med Res 2005;10:411–418.

48 Jordan E, Wagner H: Structure and properties of polysaccharides from *Viscum album* L. Oncology 1986;43(suppl 1):8–15.

49 Jung ML, Baudino S, Ribéreau-Gayon G, Beck JP: Characterization of cytotoxic proteins from mistletoe (*Viscum album* L.). Cancer Lett 1990;51:103–108.

50 Büssing A: Biological and pharmacological properties of *Viscum album* L; in Büssing A (ed): Mistletoe-The genus *Viscum*. Amsterdam, Harwood Academic Publishers, 2000, pp 123–182.

51 Kienle GS, Albonico HU, Baars E, Hamre HJ, Zimmermann P, Kiene H: Anthroposophic medicine: an integrative medical system originating in Europe. Global Adv Health Med 2013;2:20–31.

52 Urech K, Schaller G, Jäggy C: Viscotoxins, mistletoe lectins and their isoforms in mistletoe (*Viscum album* L.) extracts Iscador – analytical results on pharmaceutical processing of mistletoe. Arzneim Forsch./ Drug Res 2006;56:428–434.

53 Samuelsson G: Mistletoe toxins. Syst Zool 1974;22: 566–569.

54 Schaller G, Urech K, Giannattasio M: Cytotoxicity of different viscotoxins and extracts from the European subspecies of *Viscum album* L. Phytother Res 1996; 10:473–477.

55 Orrù S, Scaloni A, Giannattasio M, Urech K, Pucci P, Schaller G: Amino acid sequence, S-S bridges arrangement and distribution in plant tissues of thionins from *Viscum album*. Biol Chem 1997;378:989–996.

56 Romagnoli S, Fogolari F, Catalano M, Zetta L, Schaller G, Urech K, Giannattasio M, Ragona L, Molinari H: NMR solution structure of viscotoxin C1 from *Viscum album* species *coloratum* Ohwi: toward a structure-function analysis of viscotoxins. Biochem 2003:42:12503–12510.

57 Kahle B, Debreczeni JE, Sheldrick GM, Zeeck A: Vergleichende Zytotoxizitätsstudien von Viscotoxin-Isoformen und Röntgenstruktur von Viscotoxin A3 aus Mistelextrakten; in Scheer R, Bauer R, Becker H, Fintelmann V, Kemper F, Schilcher H (Hrsg): Fortschritte in der Misteltherapie. Aktueller Stand der Forschung und klinische Anwendung. Essen, KVC Verlag, 2005, pp 83–97.

58 Franz H, Ziska P, Kindt A: Isolation and properties of three lectins from mistletoe (*Viscum album* L.). Biochem J 1981;195:481–484.

59 Krauspenhaar R, Eschenburg S, Perbandt M, Kornilov V, Konareva N, Mikailova I, Stoeva S, Wacker R, Maier T, Singh T, Mikhailov A, Voelter W, Betzel C: Crystal structure of mistletoe lectin I from *Viscum album*. Biochem Biophys Res Com 1999;257:418–424.

60 Escher P, Eiblmeier M, Hetzger I, Rennenberg H: Seasonal and spatial variation of reduced sulphur compounds in mistletoes (*Viscum album*) and the xylem sap of its hosts (*Populus euramricana* and *Abies alba*). Physiol Plant 2003;117:72–78.

61 Vester F, Mai W: Zur Kenntnis der Inhaltsstoffe von *Viscum album*. I. Freie Aminosäuren. Hoppe-Seyler's Z Physiol Chem 1960;322:273–277.

62 Urech K: Accumulation of arginine in *Viscum album* L.-Seasonal variations and host dependency. J Plant Physiol 1997;151:1–5.

63 Winterfeld K, Kronenthaler A: Zur Chemie des blutdrucksenkenden Bestandteiles der Mistel (*Viscum album*). III. Mitteilung. Arch Pharm 1942;280:103–115.

64 Krzaczek T: Pharmacobotanical research on the subspecies of *Viscum album* L. IV. Acids and amines. Ann Univ M Curie Lublin Polonia 1977;32:281–291.

65 Wagner H, Jordan E: An immunologically active arabinogalactan from *Viscum album* 'berries'. Phytochemistry 1988;27:2511–2517.

66 Edlund U, Hensel A, Frose D, Pfüller U, Scheffler A: Polysaccharides from fresh *Viscum album* L. berry extract and their interaction with *Viscum album* Agglutinin I. Arzneim Forsch/Drug Res 2000;50:645–651.

67 Deliorman D, Orhan I: Fatty acid composition of *Viscum album* subspecies from Turkey. Chem Nat Compounds 2006;42:641–644.

68 Fukunaga T, Kajikawa I, Nishiya K, Watanabe Y, Suzuki N, Takeya K, Itokawa H: Studies on the constituents of the European mistletoe, *Viscum album* L. Chem Pharm Bull 1987;35:3292–3297.

69 Vicas SI, Rugina D, Socaciu C: Antioxidant activity of European mistletoe *(Viscum album)*; in Rao V (ed): Phytochemicals as Nutraceuticals – Global Approaches to their Role in Nutrition and Health. Rijeka/Croatia, InTech, 2012, pp 117–134.

70 Teuscher E: Viscum; in Bruchhausen F, et al (Hrsg): Hager's Handbuch der Pharmazeutischen Praxis. Berlin, Heidelberg, Springer, 1994, pp 1160–1183.

71 Wagner H, Feil B, Seligmann O, Petricic J, Kalogjera Z: Phenylpropanes and Lignanes of *Viscum album*, cardiovactive drugs V. Planta Med 1986;52:102–104.

72 Luczkiewicz M, Cisowski W, Kaiser P, Ochocka R, Piotrowski A: Comparative analysis of phenolic acids in mistletoe plants from various hosts. Acta Pol Pharm 2001;583:73–379.

73 Deliu C, Stirban M: Annual dynamics of assimilatory pigments in *Viscum album* L. and in its host plant, *Populus tremula* L. Contrib Bot Gradina Bot Univ Babes Bolyai Cluj 1976;11:243–249.

Konrad Urech, PhD
Hiscia Institute, Swiss Society for Cancer Research
Kirschweg 9
CH-4144 Arlesheim (Switzerland)
E-Mail k.urech@vfk.ch

Zänker KS, Kaveri SV (eds): Mistletoe: From Mythology to Evidence-Based Medicine.
Transl Res Biomed. Basel, Karger, 2015, vol 4, pp 24–38 (DOI: 10.1159/000375423)

The Anti-Cancer Activity of Mistletoe Preparations, as Related to Their Polyphenolic Profiles

Henning M. Schramm

Institute Hiscia, Society for Cancer Research, Arlesheim, Switzerland

Abstract

Polyphenols are known for their specific anti-cancer prevention and therapy actions. Polyphenols are contained in mistletoe extracts at rather fairly high concentrations, and each mistletoe preparation has a specific flavonoid profile. The question arises whether polyphenols may contribute to the anti-tumoural therapeutic potential of these preparations. Indeed, in experimental and clinical studies, the anti-tumoural actions of mistletoe preparations and of various polyphenols concur to a large extent; this paper reviews and evaluates this issue. What makes polyphenols so interesting in oncology? Very distinctively, they act differently on malignant and normal cells: they exert cytotoxic, anti-angiogenetic, anti-hormonal, detoxicative and anti-inflammatory effects on cancer cells and carcinomas, whereas they do not affect, or even act cytoprotectively, on normal cells. The polyphenols in mistletoe do not exert their actions in isolation but in combination with other substances of the plant, and additive or synergistic activity may occur between these substances. Besides the various ingredients of mistletoe preparations that have been studied in relation to their immunological activities, the polyphenols contained in these preparations may contribute to combating dysfunctional metabolism in cancer cells, the tumour and the patient. © 2015 S. Karger AG, Basel

Introduction

Although the polyphenols contained in various mistletoe preparations are known for their preventive and therapeutic anti-cancer activities, their putative roles in the anti-tumoural properties of these preparations have so far not been evaluated. Polyphenols, especially flavonoids, have unique and specific anti-tumoural modes of action that interfere with the metabolism of cancer cells and of the cancer-carrying organism, and this has elicited the interest of many oncology researchers [1]. More than 12,700 publications on flavonoids and cancer are listed in PubMed, underlining the amount of research already available on this issue.

Alongside the effect of mistletoe preparations on cancer-combating responses, the polyphenols contained in these preparations may likewise intervene in the metabolism of cancer cells and cancer patients. Various studies demonstrate, for example, that mistletoe preparations may attenuate deviant glucose and lipid metabolism in cancer cells and of the patient [2, 3]. The same has been found for polyphenols, suggesting they may contribute to the action of mistletoe preparations upon cancer metabolism.

Both types of substances reduce not only cancer risk but also cancer recurrence and mortality. When comparing the highest and lowest flavonoid nutritional intake groups, various studies found a reduction of risk, recurrence and mortality in breast, prostate and colorectal cancer patients that ranged from 30 to 60% [4–6]. Comparable results in breast cancer patients were obtained using long-term therapy with mistletoe preparations [7]. A further reduction in breast cancer mortality, decreasing to 91%, was observed when a diet rich in flavonoids was combined with a patient's regular physical activity [8].

The aim of this review is to assess and define the therapeutic role of flavonoids in the anti-tumoural activity of mistletoe preparations.

The Polyphenolic Constituents of Mistletoe Preparations

Depending on the mode of extraction, mistletoe preparations may contain different and varying amounts of polyphenols. Most mistletoe preparations used in cancer treatment are extracted by maceration, which transfers about one-third of the flavonoids contained in the plant [9]. Flavonoid profiles differ between European mistletoe preparations from deciduous and coniferous trees. The characteristic polyphenols of mistletoe preparations from deciduous trees include quercetin, gallic acid, and chlorogenic acid. Naringenin was not detected, but it is found in mistletoe preparations from coniferous trees, namely the pine [10–12]. Thus, although our knowledge on the composition of mistletoe preparations is somewhat limited, we consider quercetin and naringenin as the flavonoid substances chiefly responsible for the various metabolic anti-tumoural properties of their corresponding mistletoe preparations.

The polyphenol contents in *Viscum album fermentatum* (Iscador preparations) vary from 1.9 to 2.5% in a 20% concentration extract. On average, this corresponds to 380–492 μg/ml of polyphenols in a 2% (20 mg/ml) ampoule.

Metabolism, Energy Homeostasis and Cancer

Instead of gaining energy from oxidative phosphorylation like most normal cells, cancer cells get their energy mainly from aerobic glycolysis [13]. In addition to glucose metabolism, lipid metabolism is also dysfunctional in these cells. The beta-oxidation of fatty acids is reduced, or where it is increased, is uncoupled by UCP-2 to overcome the oxidative stress in cancer cells. Furthermore, the activity of fatty acid synthase is increased in many tumour cell types [14].

Oxidative stress plays an important role in the pathway of carcinogenesis, and it becomes even more important once a cell has switched to a malignant phenotype with the capacity for unlimited proliferation. In this case, the mitochondrial respiratory chain becomes dysfunctional and produces increasing amounts of reactive oxygen species (ROS) that have to be neutralised [13]. The cancer cell enhances the metabolic pathways required to neutralise ROS on the one hand and to form the metabolic intermediates necessary for the continuous proliferation of the cancer cell on the other hand.

This review will discuss how mistletoe preparations and polyphenols interfere with these metabolic pathways in a malignant cell.

Polyphenols Under Investigation in Oncology

In Asia, where, for example, the consumption of Genistein in soya products and catechins in green tea is high, much lower incidences of prostate cancer (25 times lower) and mammary cancers (10 times lower) were found in comparison to residents in western countries. These differences diminish substantially when Asians migrate to western countries and adopt a western lifestyle [15, 16]. Therefore, not only genetic factors but also specific different nutritional factors in Asia are considered to contribute to the reduced incidence of prostate and mammary cancer in these countries. Based on epidemiological studies, Doll and Peto estimated that an approximate average of 30–35% of all cancer deaths are attributable to nutrition [17, 18]. If dietary factors, including obesity and alcohol and tobacco consumption, are considered together, the percentage of cancer-related deaths that are attributable to them is as high as 60–90% [18].

These observations triggered a large number of epidemiological studies on the effect on cancer risk, recurrence and mortality from diets rich in polyphenols. Many of these studies showed not only a significant anti-tumoural activity but also varying susceptibility of the different cancer types to the diverse polyphenolic substances in the diet, a subject we will return to later.

Various in vitro and in vivo studies have demonstrated that polyphenols exert pro-apoptotic, anti-glycolytic, anti-lipogenic, anti-angiogenetic, anti-inflammatory, pro- and anti-oxidative, anti-aromatase, anti-oestrogenic and anti-fatigue activities. All of these modes of action, except the anti-oestrogenic and anti-aromatase activities, were, as we will refer to later, demonstrated using all mistletoe preparations. The possible hormonal activities of certain mistletoe preparations have so far not been investigated, and therefore this factor cannot be evaluated.

Some Distinctive Properties of Mistletoe Preparations and Polyphenols in Relation to Their Anti-Cancer Modes of Action

We focus here on the surprising and unique anti-tumoural mode of actions of polyphenols, which are distinct from most other chemotherapeutic substances applied in cancer treatment. Polyphenols interfere with the metabolism of cancer cells at various levels, whereas normal cells are unaffected by them. A majority of polyphenols, especially flavonols and flavones, exert their cytotoxic effects on the mitochondria of cancer cells via pro-oxidative properties, whereas they act as anti-oxidants at the same concentrations in normal cells [19–22]. Polyphenols exert anti-oestrogenic properties on breast cancer cells and at the same time act pro-oestrogenically on bone cells, thus stabilising them [23]. The flavonols and flavones inhibit fatty acid synthase without provoking weight loss that is potentially induced by synthetic inhibitors [24]. Various polyphenols depress the activity of enzymes in detoxification phase I, which are specifically overexpressed in cancer cells, whereas they stimulate enzymes in detoxification phase II, where lipophilic oncogenic substances are neutralised and excreted [25].

However, besides influencing the metabolism of cancer cells, polyphenols attenuate extreme metabolic conditions, such as obesity or underweight, which are risk factors for cancer.

The extent to which these distinctive properties have been demonstrated for mistletoe preparations is a question that will be discussed in the following sections.

Pro- or Anti-Oxidative Properties of Mistletoe Preparations and Polyphenols in Respect to Target Cells

The preventive effect of flavonoids on carcinogenesis is assumed to be mainly due to their anti-oxidative activity. However, in various cancer cell lines and in animal cancer studies, flavonoids, such as flavonols, catechins, and flavones, act pro-oxidatively in mitochondria and thus cause apoptosis of these malignant cells [20]. Flavonoids deplete glutathione, thioredoxin reductase and other radical scavengers such as catalase in cancer cells, resulting in an increase in hydrogen peroxide levels and consequently apoptosis [26–28]. Thus, depending on the malignancy or non-malignancy of the cell, various flavonoids show opposing properties [19–21].

Similar results are reported for mistletoe extracts and their isolated lectin ingredients. Mistletoe preparations exert non-toxic effects or even decreased toxicity, e.g. from methotrexate, in bone marrow cells, comparable to the effect of quercetin [29, 30]. In contrast, lectin isolated from total mistletoe extracts (host: apple tree) induces apoptotic death in cancer cells due to a remarkable generation of intracellular hydrogen peroxide [31, 32]. Mistletoe extracts have decreased glutathione and catalase levels and increased xanthine oxidase levels in cancer cells, thus indicating a reduced redox potential [31–33]. In contrast to this pro-oxidative effect in cancer cells, lectin isolated from mistletoe exerted anti-oxidative properties in a cell culture of non-malignant renal epithelial cells [34]. Mistletoe extract increased the catalase levels and reduced the xanthine oxidase levels in the plasma of cancer-bearing animals, thus improving the redox potential at this site; however, the redox potential was increased at the cancer site [32].

In the clinical situation, some cervical cancers have shown the capacity to broaden their redox potential, but these cancers did not cease proliferation when ROS generation was increased by irradiation [35]. Obviously, these carcinomas could cope with increased oxidative stress, whereas other carcinomas react to it by inducing apoptosis.

Various polyphenols and mistletoe preparations may cause apoptosis independent of increasing oxidative stress. In addition, the inhibition of fatty acid synthase by polyphenols and mistletoe preparations may induce apoptosis independent of ROS enhancement [36]; this is described in the corresponding section.

Oestrogenic and Anti-Oestrogenic Activities and Inhibition of Aromatase

In spite of their distinctive anti-oestrogenic and aromatase-inhibiting activities, various flavonoids stabilise the bone due to their anti-osteoclastic activity [37, 38]. Both quercetin and naringenin exert osteogenic properties. The stabilising effect of polyphenols on bone is of importance for cancer patients, as they usually suffer from osteoporosis or osteopenia due to their disease and chemotherapeutic treatments.

Mistletoe preparations showed oestrogen and progesterone receptor-binding properties [39]. The anti-oestrogenic effect of mistletoe preparations on cancer cells has not been investigated, but two studies were undertaken to study their osteogenic properties. *Viscum album fermentatum*, i.e. Iscador preparations, significantly inhibited the formation of osteoclasts [40]. In addition, extracts from *Viscum album coloratum*, which contains flavanone glycosides, significantly reduced the formation of multi-nucleated osteoclasts, suggesting an additional bone-stabilising effect. These effects were achieved at low concentrations within the μg/ml range [41].

Glucose Metabolism

The extent of glycolysis in cancer cells and their correspondingly increased requirement for glucose is correlated with their aggressiveness [42]. Hyperglycaemia may increase the risk of cancer, and high glucose levels in the blood are prognostically unfavourable for cancer patients. It is therefore not astonishing that anti-tumoural activity may ensue due to a reduction of increased glucose blood levels, thus improving the prognosis for cancer patients.

In various animal models, the glucose-lowering activity of various mistletoe preparations was demonstrated [3, 43, 44]. In folk and traditional medicine of various countries, mistletoe preparations are used as anti-diabetic drugs. The lectin content of mistletoe is not responsible for this activity because its removal from the mistletoe preparations did not diminish the glucose-lowering activity of these preparations [3]. Therefore, polyphenols may be responsible for the anti-diabetic and glucose-lowering activities contained in mistletoe preparations.

Due to increased glycolysis and ROS in carcinomas, advanced glycation end products are formed and are responsible for the general aggravation of the cancer disease. Furthermore, polyphenols extracted from a mistletoe preparation were found to exert anti-glycation properties [45].

Lipid Metabolism of Cancer Cells: Inhibition of Fatty Acid Synthase

Cancer cells demonstrate dysfunction not only of glucose metabolism, but also of lipid metabolism [14, 46, 47]. Fatty acid synthase is activated in all adenocarcinomas and in squamous cell carcinomas of the neck and head, kidney and bladder, and activated fatty acid synthase is a prerequisite for the proliferation of cancer cells at these tumour sites. Fatty acid synthase facilitates a specific lipid profile and alters the lipid rafts of cancer cell membranes that are required for continuous proliferation. Furthermore, lipids may be stored as droplets in the cancer cell. The more lipids that are stored in oestrogen receptor-negative mammary carcinoma cells, the more aggressive their behaviour [48]. If fatty acid synthase is inhibited, apoptosis is induced in the cancer cells.

The mistletoe preparation *Viscum album coloratum* inhibited fatty acid synthase [2]. This is not surprising, as the polyphenols contained in mistletoe preparations from deciduous trees are strong inhibitors of fatty acid synthase and consequently induce apoptosis of cancer cells. These polyphenols include quercetin, luteolin and kaempferol [49].

Inhibition of fatty acid synthase by polyphenols and by mistletoe preparations may be specific to cancer cells, as these substances do not provoke weight loss potentially caused by synthesised fatty acid synthase inhibitors [50, 51].

Naringenin, taxifolin and eriodictyol, polyphenols that are mainly found in mistletoe preparations from coniferous trees, showed fairly weak or no inhibition of fatty acid synthase [49, 52]. Traditionally, mistletoe preparations from pine tree are applied to cancer sites such as skin and brain, where fatty acid synthase does not play an important role.

How other aspects of lipid metabolism in a cancer patient are changed will be discussed in the section on obesity and underweight.

Action on Enzymes of Detoxification Phases I and II

The detoxification enzymes I and II play important roles both in the process of carcinogenesis and in cancer prevention, and the cancer-preventive effects of polyphenols are attributed mainly to their modulating effects on these two detoxification systems [53]. Studies have shown that polyphenols specifically modulate enzymes that are key players in carcinogenesis and cancers, but not in normal cells [54].

Three studies using mistletoe preparations to modulate phase I detoxification enzymes focused on those enzymes, which play a major role in normal cells but only a minor role in cancer cells. The aim of these studies was to find out whether mistletoe preparations could interfere with the metabolism of cytochrome P450 (CYP) enzymes of normal cells, resulting in possible interference with the metabolism of certain drugs. In vitro, a preparation of mistletoe harvested from fir inhibited CYP3A4, the major phase I drug-metabolising enzyme, by 20–24%, a degree considered not relevant for normal cells [55, 56]. The CYP3A4 enzyme is overexpressed in oestrogen receptor (ER)- and progesterone receptor-positive breast cancer cells. In an animal study, mistletoe extract decreased the CYP27A1 level [57]. CYP27A1 is upregulated in breast and prostate cancer cells and is a key enzyme of vitamin D3 and cholesterol metabolism.

Activation of phase II detoxifying enzymes involves various transferases and reductases. In a murine hepatoma cell line, flavonols such as quercetin were inducers of quinine reductase, whereas flavanones such as naringenin were not. In a healthy liver, as well as in hepatocarcinoma, naringenin increased glutathione S-transferase activity, whereas quercetin activity was not significantly changed.

The activation of these two phase II enzymes by mistletoe preparations has been investigated by Vicas et al. In a human ovarian carcinoma cell line, a preparation of mistletoe harvested from an apple host tree significantly increased the activity of glutathione S-transferase and quinine reductase in the cytosol [58].

Risk Factors of Cancer: Obesity and Underweight

Obesity and underweight are risk factors for cancer and prognosis factors for its further progression. In general, obese subjects have a higher risk for adenocarcinomas, especially colorectal cancer in male subjects, than persons of normal weight. In contrast, lean subjects have a higher risk of squamous cell carcinomas. Underweight and smoking specifically increase the risk for squamous cell carcinomas of the lung and oral cavity. In regards to mammary carcinoma, underweight, premenopausal women have a higher risk of developing breast cancer (oestrogen receptor-positive cancer) [59, 60]. In contrast, obese premenopausal women have a lower breast cancer risk than underweight or normal-weight patients [61]. On the other hand, overweight, postmenopausal women have a higher risk of developing oestrogen receptor-positive mammary cancer [61].

Experimental and epidemiological studies demonstrate that polyphenols may influence these metabolic risk factors. In animal and human studies, catechins and quercetin reduced the respiratory quotient, thus indicating a switch in energy fuelling from glucose to fatty acids [62, 63]. This counteracts obesity and may reduce the patient's excess weight. As already mentioned, the Korean mistletoe mimics caloric restriction, resulting in the activation of sirtuin 1 (SIRT1), thus enhancing beta-oxidation of fatty acids. Therefore, we hypothesise that mistletoe preparations from deciduous trees may activate SIRT1 in obese patients.

Corresponding results were observed after *Viscum album fermentatum Quercus* (Iscador Qu) application to diabetic rats, which decreased the concentration of cholesterol by 21.3% and of triglycerides by 32.6% in comparison to control animals [64]. These results were confirmed by other studies using mistletoe preparations [65].

Naringenin acts differently and may develop its metabolic activity more appropriately in underweight patients. Naringenin does not activate SIRT1 protein deacetylase, and consequently, no general change in energy fuelling towards beta-oxidation takes place, nor does it affect adipocytes [66, 67]. Naringenin activity is focused on liver metabolism and induces a significant reduction of total triglycerides, free fatty acids, phospholipids and cholesterol in the liver, resulting in lower plasma levels [68]. Naringenin also inhibits very-low-density lipoprotein secretion in hepatocytes [69]. Very-low-density lipoprotein levels are high in underweight premenopausal women under the influence of high oestrogen levels [59, 60].

Traditionally and based on clinical experiences, mistletoe preparations from a pine host tree that specifically contain naringenin are recommended for underweight, asthenic cancer patients, e.g. for lean premenopausal women with oestrogen receptor-positive mammary carcinomas [70]. In cachectic cancer patients, treatment with mistletoe preparations from coniferous trees resulted in attenuation of cachexia [71].

The Different Modes of Action of the Various Polyphenols and the Assumed Corresponding Actions of Mistletoe Preparations

A large number of epidemiological studies have demonstrated that polyphenolic compounds may vary in their carcinogenesis-inhibiting and anti-tumoural activities depending on the organ, histological type of cancer and hormonal sensitivity, e.g. whether the breast cancer is hormonal-responsive or not. In various studies, the uptake of specific polyphenols via diet, but not the total quantity of polyphenols, was the factor that decreased cancer incidence and cancer recurrence [72]. Epidemiological studies revealed an inverse relationship between a high nutritional uptake of flavonols, e.g. quercetin, and flavones, e.g. apigenin and luteolin, and the incidence of colorectal cancer, breast cancer in postmenopausal women and lung cancer (adenocarcinoma) [73]. One study showed an inverse relationship between the intake of apples rich in quercetin over a longer period of time and the incidence of lung cancer

of these non-nutritional studies are therefore not comparable to those from the clinical application of mistletoe preparations but merely facilitate the development of certain hypotheses related to the mode of action of polyphenols and mistletoe preparations.

Not all epidemiological studies have come to the same conclusions. This is not surprising since there are many confounding factors. If an epidemiological study, for instance, takes into account only the whole fruit intake and does not define specific fruits, it may arrive at a different conclusion from those studies where a differentiation was made between citrus fruits rich in naringenin and those rich in quercetin, such as apple or plum. An epidemiological study showed that physical activity can influence the results and, as an additive effect, can play a decisive role in reducing cancer recurrence. Many other causes, such as cancer-target specificity, inter-individual differences in absorption and metabolism, mixtures of antagonistic or synergistic agents, fat uptake, meat consumption, body mass index, menopausal status or hormonal responsiveness, may influence and cause inconsistencies in the results.

This review focuses solely on the polyphenolic content of mistletoe preparations alone. Besides these compounds, there are many other ingredients in mistletoe preparations with proven pharmacological activities, such as lectins and viscotoxins, which are known to exert pharmacological activities at nanogram levels. Additive or even synergistic effects between polyphenols and these substance classes may be possible and may lead to a hypothesis explaining the therapeutic results of mistletoe preparations subcutaneously applied to cancer patients. For instance lectin-induced apoptosis in cancer cells through increased oxidative stress is caused by certain flavonoids, such as flavonols and flavones; therefore, an additive effect between them can be hypothesised. Various researchers assume that polyphenols act additively or synergistically depending on the mixture and on other food ingredients. This assumption may explain the discrepancy between results from clinical studies that used single polyphenols and those obtained in epidemiological studies, where they were taken up in a complex mixture of substances [25].

The role of mistletoe preparations and polyphenols in the treatment of cancer may be compared with the chemopreventive agent tamoxifen. Tamoxifen has proven to inhibit early stages of carcinogenesis in experimental studies and to reduce cancer recurrence in the adjuvant situation of oestrogen receptor-positive breast cancer patients. The active metabolite of tamoxifen, endoxifen, resembles certain polyphenols, especially those with oestrogenic/anti-oestrogenic activity, and it acts synergistically with polyphenolic compounds and anti-cancer chemotherapeutics [91]. Polyphenols and mistletoe preparations seem to share the specific and non-toxic inhibition of early carcinogenesis as well as the reduction of cancer recurrence and mortality with tamoxifen. However, the various polyphenolic substances discussed here as well as the various mistletoe preparations exert a far broader scope of anti-tumoural actions on various tumour types than the agents limited to oestrogenic-responsive carcinomas.

The weak point of our review is the classification of quercetin and naringenin as lead substances in mistletoe preparations from deciduous or coniferous host trees. So far, these substances have been identified in these mistletoe preparations but have only been roughly quantified. In addition, a comprehensive analysis of all flavonoids detectable in mistletoe preparations, especially isoflavones, is lacking. Thus, it is quite possible that future analytical results will necessitate correction, changes or modification to this study's interpretation. However, as the interpretations given in this review concur with the traditional use of the different mistletoe preparations and with the pharmacological studies in animals, we assume them to be largely valid.

So far, most studies on the anti-tumoural activities of mistletoe preparations have focused on immunological activities, especially their lectins. Broadening the scope of research to the polyphenols of mistletoe preparations may open new perspectives on their anti-tumoural activity as well as their targeted application on a scientific basis.

References

1 Carocho M, Ferreira IC: The role of phenolic compounds in the fight against cancer–a review. Anticancer Agents Med Chem 2013;13:1236–1258.

2 Jung HY, Kim YH, Kim IB, Jeong JS, Lee JH, Do MS, Jung SP, Kim KS, Kim KT, Kim JB: The Korean Mistletoe (Viscum album coloratum) extract has an anti-obesity effect and protects against hepatic steatosis in mice with high-fat diet-induced obesity. Evid Based Complement Alternat Med 2013;2013:168207.

3 Gray AM, Flatt PR: Insulin-secreting activity of the traditional antidiabetic plant Viscum album (mistletoe). J Endocrinol 1999;160:409–414.

4 Hoensch H, Groh B, Edler L, Kirch W: Prospective cohort comparison of flavonoid treatment in patients with resected colorectal cancer to prevent recurrence. World J Gastroenterol 2008;14:2187–2193.

5 Guha N, Kwan ML, Quesenberry CP Jr, Weltzien EK, Castillo AL, Caan BJ: Soy isoflavones and risk of cancer recurrence in a cohort of breast cancer survivors: the Life After Cancer Epidemiology study. Breast Cancer Res Treat 2009;118:395–405.

6 Lazarevic B, Boezelijn G, Diep LM, Kvernrod K, Ogren O, Ramberg H, Moen A, Wessel N, Berg RE, Egge-Jacobsen W, Hammarstrom C, Svindland A, Kucuk O, Saatcioglu F, Tasken KA, Karlsen SJ: Efficacy and safety of short-term genistein intervention in patients with localized prostate cancer prior to radical prostatectomy: a randomized, placebo-controlled, double-blind Phase 2 clinical trial. Nutr Cancer 2011;63:889–898.

7 Grossarth-Maticek R, Ziegler R: Prospective controlled cohort studies on long-term therapy of breast cancer patients with a mistletoe preparation (Iscador). Forsch Komplementmed 2006;13:285–292.

8 George SM, Irwin ML, Smith AW, Neuhouser ML, Reedy J, McTiernan A, Alfano CM, Bernstein L, Ulrich CM, Baumgartner KB, Moore SC, Albanes D, Mayne ST, Gail MH, Ballard-Barbash R: Postdiagnosis diet quality, the combination of diet quality and recreational physical activity, and prognosis after early-stage breast cancer. Cancer Causes Control 2011;22:589–598.

9 Jager S, Beffert M, Hoppe K, Nadberezny D, Frank B, Scheffler A: Preparation of herbal tea as infusion or by maceration at room temperature using mistletoe tea as an example. Sci Pharm 2011;79:145–155.

10 Lorch E: New investigations on flavonoids from viscum album L. ssp abietis, album and austriacum. Z Naturforsch 1993;48:105–107.

11 Haas K, Bauer M, Wollenweber E: Cuticular waxes and flavonol aglycones of mistletoes. Z Naturforsch C 2003;58:464–470.

12 Vicas SI, Rugina OD, Leopold L, Pintea A, Socaciu C: HPLC fingerprint of bioactive compounds and antioxidant activities of viscum album from different host trees. Not Bot Hort Agrobot Cluj 2011;39:48–57.

13 Seyfried TN, Shelton LM: Cancer as a metabolic disease. Nutr Metab (Lond) 2010;7:7.

14 Baenke F, Peck B, Miess H, Schulze A: Hooked on fat: the role of lipid synthesis in cancer metabolism and tumour development. Dis Model Mech 2013;6:1353–1363.

15 Kimura T: East meets West: ethnic differences in prostate cancer epidemiology between East Asians and Caucasians. Chin J Cancer 2012;31:421–429.

16 Ziegler RG, Hoover RN, Pike MC, Hildesheim A, Nomura AM, West DW, Wu-Williams AH, Kolonel LN, Horn-Ross PL, Rosenthal JF, Hyer MB: Migration patterns and breast cancer risk in Asian-American women. J Natl Cancer Inst 1993;85:1819–1827.

17 Doll R, Peto R: The causes of cancer: quantitative estimates of avoidable risks of cancer in the United States today. J Natl Cancer Inst 1981;66:1191–1308.

18 Anand P, Kunnumakkara AB, Sundaram C, Harikumar KB, Tharakan ST, Lai OS, Sung B, Aggarwal BB: Cancer is a preventable disease that requires major lifestyle changes. Pharm Res 2008;25:2097–2116.

19 Babich H, Selevan AR, Ravkin ER: Glutathione as a mediator of the in vitro cytotoxicity of a green tea polyphenol extract. Toxicol Mech Methods 2007;17: 357–369.

20 Hui C, Yujie F, Lijia Y, Long Y, Hongxia X, Yong Z, Jundong Z, Qianyong Z, Mantian M: MicroRNA-34a and microRNA-21 play roles in the chemopreventive effects of 3,6-dihydroxyflavone on 1-methyl-1-nitrosourea-induced breast carcinogenesis. Breast Cancer Res 2012;14:R80.

21 Yamamoto T, Lewis J, Wataha J, Dickinson D, Singh B, Bollag WB, Ueta E, Osaki T, Athar M, Schuster G, Hsu S: Roles of catalase and hydrogen peroxide in green tea polyphenol-induced chemopreventive effects. J Pharmacol Exp Ther 2004;308:317–323.

22 Gibellini L, Pinti M, Nasi M, De Biasi S, Roat E, Bertoncelli L, Cossarizza A: Interfering with ROS metabolism in cancer cells: the potential role of quercetin. Cancers (Basel) 2010;2:1288–1311.

23 Veprik A, Khanin M, Linnewiel-Hermoni K, Danilenko M, Levy J, Sharoni Y: Polyphenols, isothiocyanates, and carotenoid derivatives enhance estrogenic activity in bone cells but inhibit it in breast cancer cells. Am J Physiol Endocrinol Metab 2012; 303:E815–E824.

24 Puig T, Turrado C, Benhamu B, Aguilar H, Relat J, Ortega-Gutierrez S, Casals G, Marrero PF, Urruticoechea A, Haro D, Lopez-Rodriguez ML, Colomer R: Novel inhibitors of fatty acid synthase with anticancer activity. Clin Cancer Res 2009;15:7608–7615.

25 Liu RH: Potential synergy of phytochemicals in cancer prevention: mechanism of action. J Nutr 2004; 134(suppl):3479S–3485S.

26 Kachadourian R, Day BJ: Flavonoid-induced glutathione depletion: potential implications for cancer treatment. Free Radic Biol Med 2006;41:65–76.

27 Lu J, Papp LV, Fang J, Rodriguez-Nieto S, Zhivotovsky B, Holmgren A: Inhibition of mammalian thioredoxin reductase by some flavonoids: implications for myricetin and quercetin anticancer activity. Cancer Res 2006;66:4410–4418.

28 Babich H, Schuck AG, Weisburg JH, Zuckerbraun HL: Research strategies in the study of the pro-oxidant nature of polyphenol nutraceuticals. J Toxicol 2011;2011:467305.

29 Sekeroglu ZA, Sekeroglu V: Effects of Viscum album L. extract and quercetin on methotrexate-induced cyto-genotoxicity in mouse bone-marrow cells. Mutat Res 2012;746:56–59.

30 Sekeroglu V, Aydin B, Sekeroglu ZA: Viscum album L. extract and quercetin reduce cyclophosphamide-induced cardiotoxicity, urotoxicity and genotoxicity in mice. Asian Pac J Cancer Prev 2011;12:2925–2931.

31 Kim MS, Lee J, Lee KM, Yang SH, Choi S, Chung SY, Kim TY, Jeong WH, Park R: Involvement of hydrogen peroxide in mistletoe lectin-II-induced apoptosis of myeloleukemic U937 cells. Life Sci 2003;73: 1231–1243.

32 Stan RL, Hangan AC, Dican L, Sevastre B, Hanganu D, Catoi C, Sarpataki O, Ionescu CM: Comparative study concerning mistletoe viscotoxins antitumor activity. Acta Biol Hung 2013;64:279–288.

33 Park YK, Do YR, Jang BC: Apoptosis of K562 leukemia cells by Abnobaviscum F(R), a European mistletoe extract. Oncol Rep 2012;28:2227–2232.

34 Kim BK, Choi MJ, Park KY, Cho EJ: Protective effects of Korean mistletoe lectin on radical-induced oxidative stress. Biol Pharm Bull 2010;33:1152–1158.

35 Demirci S, Ozsaran Z, Celik HA, Aras AB, Aydin HH: The interaction between antioxidant status and cervical cancer: a case control study. Tumori 2011; 97:290–295.

36 Mukherjee A, Sikdar S, Bishayee K, Boujedaini N, Khuda-Bukhsh AR: Flavonol isolated from ethanolic leaf extract of Thuja occidentalis arrests the cell cycle at G2-M and induces ROS-independent apoptosis in A549 cells, targeting nuclear DNA. Cell Prolif 2014; 47:56–71.

37 Trzeciakiewicz A, Habauzit V, Horcajada MN: When nutrition interacts with osteoblast function: molecular mechanisms of polyphenols. Nutr Res Rev 2009;22:68–81.

38 Wu YW, Chen SC, Lai WF, Chen YC, Tsai YH: Screening of flavonoids for effective osteoclastogenesis suppression. Anal Biochem 2013;433:48–55.

39 Zava DT, Dollbaum CM, Blen M: Estrogen and progestin bioactivity of foods, herbs, and spices. Proc Soc Exp Biol Med 1998;217:369–378.

40 Kübler S: Einfluss von Mistelextrakten auf die Differenzierung von RAW 264.7 Makropahgen zu Osteoklasten; thesis, Zürcher Hochschule für Angewandte Wissenschaften, 2012.

41 Han N, Huang T, Wang YC, Yin J, Kadota S: Flavanone glycosides from viscum coloratum and their inhibitory effects on osteoclast formation. Chem Biodivers 2011;8:1682–1688.

42 Gatenby RA, Gillies RJ: Why do cancers have high aerobic glycolysis? Nat Rev Cancer 2004;4:891–899.

43 Eno AE, Ofem OE, Nku CO, Ani EJ, Itam EH: Stimulation of insulin secretion by Viscum album (mistletoe) leaf extract in streptozotocin-induced diabetic rats. Afr J Med Med Sci 2008;37:141–147.

44 Gren A, Formicki G: Effects of iscador and vincristine and 5-fluorouracil on brain, liver, and kidney element levels in alloxan-induced diabetic mice. Biol Trace Elem Res 2013;152:219–224.

45 Choudhary MI, Maher S, Begum A, Abbaskhan A, Ali S, Khan A: Characterization and antiglycation activity of phenolic constituents from Viscum album (European Mistletoe). Chem Pharm Bull (Tokyo) 2010;58:980–982.

46 Schug ZT, Frezza C, Galbraith LC, Gottlieb E: The music of lipids: how lipid composition orchestrates cellular behaviour. Acta Oncol 2012;51:301–310.

47 Bielecka-Dabrowa A, Hannam S, Rysz J, Banach M: Malignancy-associated dyslipidemia. Open Cardiovasc Med J 2011;5:35–40.

48 Antalis CJ, Uchida A, Buhman KK, Siddiqui RA: Migration of MDA-MB-231 breast cancer cells depends on the availability of exogenous lipids and cholesterol esterification. Clin Exp Metastasis 2011;28: 733–741.

49 Brusselmans K, Vrolix R, Verhoeven G, Swinnen JV: Induction of cancer cell apoptosis by flavonoids is associated with their ability to inhibit fatty acid synthase activity. J Biol Chem 2005;280:5636–5645.

50 Flavin R, Peluso S, Nguyen PL, Loda M: Fatty acid synthase as a potential therapeutic target in cancer. Future Oncol 2010;6:551–562.

51 Puig T, Vazquez-Martin A, Relat J, Petriz J, Menendez JA, Porta R, Casals G, Marrero PF, Haro D, Brunet J, Colomer R: Fatty acid metabolism in breast cancer cells: differential inhibitory effects of epigallocatechin gallate (EGCG) and C75. Breast Cancer Res Treat 2008;109:471–479.

52 Yadegarynia S, Pham A, Ng A, Nguyen D, Lialiutska T, Bortolazzo A, Sivryuk V, Bremer M, White JB: Profiling flavonoid cytotoxicity in human breast cancer cell lines: determination of structure-function relationships. Nat Prod Commun 2012;7:1295–1304.

53 Androutsopoulos VP, Papakyriakou A, Vourloumis D, Tsatsakis AM, Spandidos DA: Dietary flavonoids in cancer therapy and prevention: substrates and inhibitors of cytochrome P450 CYP1 enzymes. Pharmacol Ther 2010;126:9–20.

54 Ciolino HP, Daschner PJ, Yeh GC: Dietary flavonols quercetin and kaempferol are ligands of the aryl hydrocarbon receptor that affect CYP1A1 transcription differentially. Biochem J 1999;340:715–722.

55 Engdal S, Nilsen OG: In vitro inhibition of CYP3A4 by herbal remedies frequently used by cancer patients. Phytother Res 2009;23:906–912.

56 Doehmer J, Eisenbraun J: Assessment of extracts from mistletoe (Viscum album) for herb-drug interaction by inhibition and induction of cytochrome P450 activities. Phytother Res 2012;26:11–17.

57 Culcu T, Genclar-Ozkan AM, Sen A, Adali O: The effects of folk medicinal plant Viscum album L. on protein and mRNA expressions of CYP27A1 in rat liver. FEBS Journal 2012;279:97.

58 Vicas SI, Rugina D, Sconta Z, Pintea A, Socaciu C: The in vitro antioxidant and anti-proliferative effect and induction of pase II enzymes by mistletoe (Viscum Album) extract. UASVM Agriculture 2011;68:482–491.

59 Gerber M: Re: 'Body size and breast cancer risk among women under age 45 years'. Am J Epidemiol 1997;145:669–670.

60 Schaefer EJ, Foster DM, Zech LA, Lindgren FT, Brewer HB Jr, Levy RI: The effects of estrogen administration on plasma lipoprotein metabolism in premenopausal females. J Clin Endocrinol Metab 1983;57:262–267.

61 Vrieling A, Buck K, Kaaks R, Chang-Claude J: Adult weight gain in relation to breast cancer risk by estrogen and progesterone receptor status: a meta-analysis. Breast Cancer Res Treat 2010;123:641–649.

62 Dulloo AG, Duret C, Rohrer D, Girardier L, Mensi N, Fathi M, Chantre P, Vandermander J: Efficacy of a green tea extract rich in catechin polyphenols and caffeine in increasing 24-h energy expenditure and fat oxidation in humans. Am J Clin Nutr 1999;70: 1040–1045.

63 Koch CE, Ganjam GK, Steger J, Legler K, Stohr S, Schumacher D, Hoggard N, Heldmaier G, Tups A: The dietary flavonoids naringenin and quercetin acutely impair glucose metabolism in rodents possibly via inhibition of hypothalamic insulin signalling. Br J Nutr 2013;109:1040–1051.

64 Gren A, Formicki G, Stawarz R, Matiniakova M, Omelka R: Effect of Iscador on selected parameters of the metabolic block in the animal type Diabetes induced by Alloksan. Journal of Microbiology, Biotechnology and Food Sciences 2011;1:215–224.

65 Avci G, Kupeli E, Eryavuz A, Yesilada E, Kucukkurt I: Antihypercholesterolaemic and antioxidant activity assessment of some plants used as remedy in Turkish folk medicine. J Ethnopharmacol 2006;107:418–423.

66 Morikawa K, Nonaka M, Mochizuki H, Handa K, Hanada H, Hirota K: Naringenin and hesperetin induce growth arrest, apoptosis, and cytoplasmic fat deposit in human preadipocytes. J Agric Food Chem 2008;56:11030–11037.

67 Zygmunt K, Faubert B, MacNeil J, Tsiani E: Naringenin, a citrus flavonoid, increases muscle cell glucose uptake via AMPK. Biochem Biophys Res Commun 2010;398:178–183.

68 Huong DT, Takahashi Y, Ide T: Activity and mRNA levels of enzymes involved in hepatic fatty acid oxidation in mice fed citrus flavonoids. Nutrition 2006; 22:546–552.

69 Mulvihill EE, Allister EM, Sutherland BG, Telford DE, Sawyez CG, Edwards JY, Markle JM, Hegele RA, Huff MW: Naringenin prevents dyslipidemia, apolipoprotein B overproduction, and hyperinsulinemia in LDL receptor-null mice with diet-induced insulin resistance. Diabetes 2009;58:2198–2210.

70 Debus M: Die Mistel in der Krebstherapie. Der Merkurstab 2014;67:55–61.

71 Girke M: Innere Medizin: Grundlagen und Therapeutische Konzepte der Anthroposophischen Medizin. Berlin, Salumed Verlag, 2010.

72 Hertog MG, Feskens EJ, Hollman PC, Katan MB, Kromhout D: Dietary flavonoids and cancer risk in the Zutphen Elderly Study. Nutr Cancer 1994;22:175–184.

73 Woo HD, Kim J: Dietary flavonoid intake and risk of stomach and colorectal cancer. World J Gastroenterol 2013;19:1011–1019.

74 Knekt P, Jarvinen R, Seppanen R, Hellovaara M, Teppo L, Pukkala E, Aromaa A: Dietary flavonoids and the risk of lung cancer and other malignant neoplasms. Am J Epidemiol 1997;146:223–230.

75 Rossi M, Garavello W, Talamini R, Negri E, Bosetti C, Dal Maso L, Lagiou P, Tavani A, Polesel J, Barzan L, Ramazzotti V, Franceschi S, La Vecchia C: Flavonoids and the risk of oral and pharyngeal cancer: a case-control study from Italy. Cancer Epidemiol Biomarkers Prev 2007;16:1621–1625.

76 Lam TK, Shao S, Zhao Y, Marincola F, Pesatori A, Bertazzi PA, Caporaso NE, Wang E, Landi MT: Influence of quercetin-rich food intake on microRNA expression in lung cancer tissues. Cancer Epidemiol Biomarkers Prev 2012;21:2176–2184.

77 Christensen KY, Naidu A, Parent ME, Pintos J, Abrahamowicz M, Siemiatycki J, Koushik A: The risk of lung cancer related to dietary intake of flavonoids. Nutr Cancer 2012;64:964–974.

78 Hakim IA, Harris RB: Joint effects of citrus peel use and black tea intake on the risk of squamous cell carcinoma of the skin. BMC Dermatol 2001;1:3.

79 Garavello W, Rossi M, McLaughlin JK, Bosetti C, Negri E, Lagiou P, Talamini R, Franceschi S, Parpinel M, Dal Maso L, La Vecchia C: Flavonoids and laryngeal cancer risk in Italy. Ann Oncol 2007;18:1104–1109.

80 Rossi M, Garavello W, Talamini R, La Vecchia C, Franceschi S, Lagiou P, Zambon P, Dal Maso L, Bosetti C, Negri E: Flavonoids and risk of squamous cell esophageal cancer. Int J Cancer 2007;120:1560–1564.

81 Foschi R, Pelucchi C, Dal Maso L, Rossi M, Levi F, Talamini R, Bosetti C, Negri E, Serraino D, Giacosa A, Franceschi S, La Vecchia C: Citrus fruit and cancer risk in a network of case-control studies. Cancer Causes Control 2010;21:237–242.

82 Hui C, Qi X, Qianyong Z, Xiaoli P, Jundong Z, Mantian M: Flavonoids, flavonoid subclasses and breast cancer risk: a meta-analysis of epidemiologic studies. PLoS One 2013;8:c54318.

83 Torres-Sanchez L, Galvan-Portillo M, Wolff MS, Lopez-Carrillo L: Dietary consumption of phytochemicals and breast cancer risk in Mexican women. Public Health Nutr 2009;12:825–831.

84 Malin AS, Qi D, Shu XO, Gao YT, Friedmann JM, Jin F, Zheng W: Intake of fruits, vegetables and selected micronutrients in relation to the risk of breast cancer. Int J Cancer 2003;105:413–418.

85 de Vries JH, Hollman PC, Meyboom S, Buysman MN, Zock PL, van Staveren WA, Katan MB: Plasma concentrations and urinary excretion of the antioxidant flavonols quercetin and kaempferol as biomarkers for dietary intake. Am J Clin Nutr 1998;68:60–65.

86 Hollman PC, vd Gaag M, Mengelers MJ, van Trijp JM, de Vries JH, Katan MB: Absorption and disposition kinetics of the dietary antioxidant quercetin in man. Free Radic Biol Med 1996;21:703–707.

87 Hollman PC, van Trijp JM, Buysman MN, van der Gaag MS, Mengelers MJ, de Vries JH, Katan MB: Relative bioavailability of the antioxidant flavonoid quercetin from various foods in man. FEBS Lett 1997;418:152–156.

88 Bieger J, Cermak R, Blank R, de Boer VC, Hollman PC, Kamphues J, Wolffram S: Tissue distribution of quercetin in pigs after long-term dietary supplementation. J Nutr 2008;138:1417–1420.

89 So FV, Guthrie N, Chambers AF, Moussa M, Carroll KK: Inhibition of human breast cancer cell proliferation and delay of mammary tumorigenesis by flavonoids and citrus juices. Nutr Cancer 1996;26:167–681.

90 Knopfl-Sidler F, Viviani A, Rist L, Hensel A: Human cancer cells exhibit in vitro individual receptiveness towards different mistletoe extracts. Pharmazie 2005;60:448–454.

91 Lewandowska U, Gorlach S, Owczarek K, Hrabec E, Szewczyk K: Synergistic interactions between anticancer chemotherapeutics and phenolic compounds and anticancer synergy between polyphenols. Postepy Hig Med Dosw (Online) 2014;68:528–540.

Henning M. Schramm, DVM
Institute Hiscia
Society for Cancer Research
Kirschweg 9, CH–4144 Arlesheim (Switzerland)
E-Mail h.schramm@vfk.ch

Zänker KS, Kaveri SV (eds): Mistletoe: From Mythology to Evidence-Based Medicine.
Transl Res Biomed. Basel, Karger, 2015, vol 4, pp 39–47 (DOI: 10.1159/000375424)

From Berlin and Witten to Southampton and Hamburg: 25 Years of Mistletoe Research Cooperation

Udo Schumacher · Uwe Pfüller

Institute of Anatomy and Experimental Morphology, University Cancer Center Hamburg, University Medical Center Hamburg-Eppendorf, Hamburg, Germany

Abstract

A summary of the joint projects of the Witten and Southampton/Hamburg collaboration on mistletoe lectin (ML) research is given. The Witten group has isolated all three major MLs, and the other group has investigated the MLs' biological action on tumor cells in vitro and in vivo. The main conclusion of this work is that although MLs are cytotoxic to tumor cells in vitro, their main anti-tumor action in vivo is mediated via immunological mechanisms. In addition, a number of histochemical studies were conducted with MLs. The ability of ML-I to label microglial cells is a prominent feature of MLs in histochemical applications. Finally, a tabular description of the isolation of the three MLs is given.

Mistletoe Research and International Meetings

The 6th Interlec Meeting, which took place from September 2 to 6, 1984, in Posznan, Poland, was organized by Thorkild Bog-Hansen and Jan Breborowicz and brought me (US) in contact with Hartmut Franz, who was the director of the Staatliche Institut für Immunpräparate and Nährmedien in East Berlin at that time. It was from him that I first heard about mistletoe lectins (MLs). My discussions with him deepened during the next Interlec meeting in Brussels from August 18 to 23, 1985, and at the 9th Interlec Meeting, held in Cambridge from July 27 to 31, 1987. Prof. Franz and one of the authors (UP) organized the 13th Interlec Meeting in Berlin in 1991, where one of the sessions was dedicated to MLs. Shortly afterward, the Staatliche Institut für Immunpräparate and Nährmedien group went to the University of Witten/

Herdecke, and us became involved in the group when they consulted him concerning the future of the group after the untimely death of Prof. Franz on May 2, 1992. It was agreed that Uwe Pfüller should take over the group and that we would start a close collaboration that has lasted until today. Having retired from Witten, Uwe is now working within our group in Hamburg. Over the years, the professional relationship has developed into a personal friendship. The following account of our joint work will focus on the use of MLs in histochemistry in nononcological applications and in both histochemical and functional investigations of MLs in oncology. The broadness of this spectrum of research questions is due to the fact that MLs are very versatile tools.

The Lectin Story

Lectins are by definition carbohydrate-binding proteins of nonimmunological origin that have at least two carbohydrate-binding sites. Hence, they can bind to the carbohydrate residues of erythrocytes and can crosslink them, which is called agglutination. Therefore, lectins are often alternatively called agglutinins, as they were initially identified based on their ability to agglutinate red blood cells. To understand why MLs are of such interest in medicine, one has to understand the role of carbohydrate residues in cell biology. Carbohydrate residues form the outermost layer of the cell membrane, or the glycocalyx. Because of this exposed position, the terminal carbohydrate residues of the glycocalyx, which can be recognized by lectins, play an important functional role in cell-to-cell and cell-to-matrix interactions. For example, several cell populations can be distinguished from each other by their carbohydrate coat, which in turn can be demonstrated based on its lectin-binding pattern. The carbohydrate residues in this coat can change during malignant progression, which is one hallmark of cancer. Therefore, detection of the carbohydrate residues of cells makes lectins an attractive tool to analyze carbohydrate residues of cells and tissues. As carbohydrate residues have the additional advantage of not being very influenced by formalin fixation and paraffin embedding, lectins can therefore be easily applied to archival material.

MLs are additionally attractive lectins in biomedical research because they are toxic lectins. Their toxicity is due to the fact that MLs consist of two protein chains linked by a disulfide bond. The ML A-chain is a toxic protein that inactivates ribosomes at an astonishing rate in living cells (1 A-chain molecule can kill one cell and is linked to the carbohydrate-binding B-chain via one interchain disulfide bond). Three different toxic MLs have been isolated, designated as ML-I, ML-II and ML-III. Their main difference is the carbohydrate specificity of the B-chain. ML-I mainly binds to galactose; ML-II, to both galactose (Gal) and N-acetylgalactosamine (GalNAc); and ML-III, to GalNAc alone [1]. Because of their subtly different carbohydrate specificities, MLs are excellent tools for differentiating between the Gal/

Schumacher · Pfüller

GalNAc carbohydrate residues of cells and tissues. Having provided this brief introduction, we will now describe related research. While antibodies are often species specific, lectins are carbohydrate specific and can therefore easily be examined in our joint investigations.

As I [US] am primarily a lectin histochemist, our first joint paper investigated the use of ML-I in Alzheimer's disease. Together with Japanese neuropathologists, Hartmut Franz, who had also worked in pathology for a number of years, had already described ML-I as binding specifically to microglial cells in the normal human and rodent brains [2]. We extended his investigations to the pathological human brain by looking at microglial cells in the center of amyloid plaques in the brain of Alzheimer's patients [3]. In addition to endothelial and microglial cells themselves, ML-I also labeled a proportion of the plaque proteins. As the glycosylation detected by ML-I is very cell specific, it has to be presumed that glycoproteins secreted from the microglial cells were associated with the deposited β-amyloid within the plaques and stained by ML-I lectin histochemistry. Next, we investigated the binding pattern of all three MLs with respect to their ability to label microglial cells and plaque proteins. The results of this study showed that labeling of microglial cells and plaque proteins is particular to ML-I, while ML-II and ML-III do not label microglial cells or plaque proteins [4].

As mentioned above, lectins can be used to characterize particular cell populations. Because of their flat and elongated shape, endothelial cells, and especially those in microvessels, are often hard to identify in tissue sections, especially if the lumen of the capillary has collapsed. Therefore, histochemical markers for endothelial cells are in high demand. In one study, we investigated the lectin-binding sites of quail and chicken endothelial cells [5]. These two avian species are widely used in embryological research, as cells from both species do not show transplant rejection, so the fate of cells transplanted during development can be traced because their nuclei show different chromatin patterns. All three MLs bound to endothelial cells in both species, while *Wisteria floribunda* agglutinin bound to quail endothelium both in embryonic and adult tissues. By combining the lectins from these two species, tracing of endothelial cells in embryological studies should become possible.

Lectins and Immunocompetent Cells

Another area of morphological research in which cell identification is difficult is in the lymphatic system, where most of the different cell populations cannot be distinguished using classical histological stains. We therefore used lectins to decorate cell populations in chicken, as antibodies against avian immune cells that could be applied to formalin-fixed paraffin-embedded tissue sections were not widely available at that time. We could show that several lectins bound to lymphocytes and macrophages in immune organs (thymus, spleen, bursa of Fabricius, bone marrow) in chicken. However, MLs were not particularly well suited to labeling lymphocytes and monocytes but bound to epithelial reticular cells in the thymus and to the follicle-associated epithelium in the bursa of Fabricius [6]. A subsequent study investigated the gut epithelial lining and its associated lymphoid system in greater detail. Focusing on M-cells, we could show that the lectin-binding pattern of M-cells varied according to the section of the gut, so no universal M-cell marker was observed in chicken, in contrast to mammals [7].

Lectins and Malignant Tumors

The main focus of our joint research, however, has been cancer research. Again, we started with lectin histochemistry of malignant tumors. This subject area got off the ground following two landmark studies by Anthony Leathem and Susan Brooks, who showed that binding of the GalNAc-specific lectin *Helix pomatia* agglutinin (HPA) to breast cancer cells in tissue sections was associated with a poor prognosis for the patient [8, 9]. This study was later extended to colon cancer [10]. We therefore extended this study further and applied Gal-selective ML-I and GalNAc-selective ML-III to colon sections from colon cancer patients from the previous study. Binding of the two MLs to the colon cancer cells was not of prognostic significance, indicating that the GalNAc carbohydrate residue detected by HPA is different from that recognized by ML-III, which has the same nominal carbohydrate specificity for GalNAc as HPA does [11]. This is another good example of the ability of lectins to recognize very subtle differences in the conformation of terminal carbohydrate residues.

In yet another study, we could show that in addition to breast and colon cancer, HPA is also a prognostic indicator in malignant melanoma [12]. We therefore also tested the three MLs for their ability to predict metastasis formation and hence prognosis in melanoma patients. Only very intense binding of ML-I was associated with metastasis formation in our patient cohort, indicating that very dense decoration of melanoma cells is associated with metastasis formation [13]. A later study showed that ML-I is not suitable for sentinel node diagnostics, as it lacks sensitivity [14].

As many ML-I binding sites were detected on metastasizing melanomas and as metastasized melanomas are the prime target of systemic ML therapy, we investigated the cytotoxic action of MLs on melanoma cells [15]. We found that all melanoma cell lines tested became apoptotic when treated with all three MLs. Hence, all three MLs are primarily cytotoxic. As MLs are primarily cytotoxic not only to melanoma cells but also to immunocompetent cells, the mode of action of MLs on melanoma cells may also depend on cytokines that are released during ML-induced apoptosis of immunocompetent cells. Hence, we also tested the effects of the most commonly released cytokines after ML application to immunocompetent cells, namely,

TNF-α, IL-1, and IL-6. Again, we could not show any growth-stimulatory effect of these cytokines on melanoma cells. Hence, according to our investigations, MLs and the cytokines released by them are not growth stimulatory, as proposed by Gabius et al. [16].

In these investigations, one cell line stood out due to its sensitivity to ML-I treatment: the human melanoma cell line MV3 was particularly sensitive to ML-I. We therefore thought that this cell line would be particularly well suited for investigation of the cytotoxic effects of ML-I in vivo. However, to our great amazement, the major effects of ML-I in this particularly ML-I-sensitive cell line was caused by immunological effects, and not by direct cytotoxic ones [17]. At 30 ng ML-I per kg body weight of a mouse, the weight of the primary tumor was decreased by 35% compared with the control, and the number of spontaneous lung metastases decreased by 55%. Higher doses of ML-I (150 and 500 ng/kg body weight) did not have these effects, which would have been expected if the effects of ML-I were indeed caused by ML-I cytotoxicity. However, no cytotoxic effect could be inferred from these results, and hence, we investigated the presence of immunocompetent cells in the primary xenografts. Using another lectin, *Bandeira simplicifolia* agglutinin I, we were able to identify tumor-infiltrating dendritic cells (DCs). ML-I treatment (50 ng/kg body weight) almost doubled the number of infiltrating DCs (35 vs. 60 DCs on average per high-power field [×400 magnification]). Using higher doses of ML-I reduced the number of DCs in the primary tumor, as ML-I induced greater apoptosis of the DCs. Hence, the therapeutic effect of ML-I can clearly be ascribed to the immunostimulatory effects of ML-I, and not to direct cytotoxic effects, despite the ML-I ultrasensitivity of the MV3 cell line used in these experiments.

Prior to our investigations in melanoma cells, we had investigated the effect of the three MLs in breast cancer cell lines [18]. Although differences in the cytotoxic effects between the cell lines and the three MLs were noted, they were overall relatively marginal, and the concentrations needed to kill the tumor cells in cell culture were above those reached in the serum of patients by ML therapy. These results therefore implied an immunological mechanism of the killing of tumor cells, as later observed in the melanoma studies cited above.

Although MLs are incorporated into the cell by phagocytosis, they must be transferred across the endosomal/lysosomal membrane, as they act as ribosome-inactivating proteins. Intracytoplasmic proteins are not degraded by lysosomes but are degraded in the proteasome, a multi-enzyme complex within the cytoplasm. However, augmentation of the cytotoxic effect of ML-I by the proteasome inhibitor bortezomib (Velcade®) was only marginal, indicating that the only very little ML-I internalized into the cytoplasm is degraded by the proteasomal pathway [19].

Another way of using MLs is to separate the toxic A-chain from the sugar-binding B-chain and to conjugate it to another targeting molecule to create, e.g., an immunotoxin using an antibody. As HPA preferentially recognized metastatic cancer cells in breast cancer, colon cancer and melanoma, we created an HPA/ML A-chain toxin and tested this conjugate in vitro. Unfortunately, the HPA binding characteristics were changed by this conjugation, and HPA lost its specificity, so this approach was deemed unsuitable and was not carried out further [20].

As aqueous mistletoe extracts are often given to terminally ill cancer patients, we investigated the effects of MLs on human HT29 colon cancer cells over-expressing the glycoprotein multi-drug resistance 1 (MDR-1) [21]. MDR-1 expression was achieved in two different ways. First, MDR was induced by increasing the concentration of a cytotoxic drug known to be transported by MDR-1. Second, MDR-1 over-expression was achieved by retroviral transduction with the *mdr-1* gene. The results clearly indicated that the MDR-1-over-expressing cells induced by rising drug concentrations were more susceptible to ML treatment than the virally transduced cells were. In addition to MDR-1 over-expression, multiple other changes take place in these cells, which include glycosylation changes that allow for higher ML sensitivity. This is, in principle, good news for cancer patients. However, how many of these glycosylation changes can be expected if MDR is induced in patients remains unclear, as the concentrations needed to induce MDR-1 over-expression are ultimately much higher than those concentrations given to patients.

Lectins and Binding Partners

MLs are among the most toxic proteins known, and as lectin-sugar interactions are often weaker in binding than protein-protein interactions are, we were always intrigued whether proteins and not glycostructures are the true binding partners of MLs. ML-I has been shown to strictly bind to alpha-2,6-sialylated neo-lacto series gangliosides [22]. However, peptides can also bind to the carbohydrate-recognizing domain of lectins and can thus mimic sugars, serving as so-called glycomimetic peptides. If glycomimetic peptides are identified, they can be sequenced, and the identified sequence can be compared with databases of human proteins in order to identify possible protein interaction partners of MLs. Using this approach, we could demonstrate that ML-I binding peptide has 100% sequence homology to the sequence of human MDR protein 5. However, as this sequence is part of the protein's transmembrane domain, it cannot act as a binding partner for ML-I in living cells [23].

Summary

Most of our investigations showed that MLs primarily act via immunological mechanisms, presumably by binding to immunocompetent cells and destroying them via MLs' cytotoxic activity. While these cells are dying, they release their immunomodulatory content, which then acts on immune cells, which in turn act

                                                    Schumacher · Pfüller

Table 1. Isolation of mistletoe lectins of the ribosome-interacting protein II type and the hevein type

Step	Procedure	Result
Isolation of ML-I, ML-II, ML-III		
Extraction (4°C)	Fresh material, homogenized: 1 part + 2 parts 0.2 M acetic acid Dried plant powder: 1 part + 5 parts 0.1 M acetic acid	
Filtration and pH adjustment	Filtration; pH to 4 using concentrated acetic acid	Crude extract
Solid phase extraction	1. Absorption: 100 ml crude extract shaken for 1 h with 1 g SP-Sephadex C-50, solid phase put into column, unbound proteins washed away with 0.2 M acetate buffer, pH 4 (buffer A-B) 2. Elution: 0.1 M Tris buffer, pH 8, 0.5 M NaCl (buffer T-B), pH controlled to 7–8	Enrichment of proteins Crude lectins
Affinity chromatography	1. Absorption: lactosyl-Sepharose 4B, equilibration with buffer T-B, protein-free washing with buffer T-B 2. Elution with buffer T-B + 0.015 M Gal 3. Elution with buffer T-B + 0.1 M Gal	Enrichment and purification of lectins Elution of ML-II/III Elution of ML-I
Precipitation (4°C)	Addition of solid ammonium sulfate to a final concentration of 3–3.2 M ammonium sulfate	Enrichment and storage of ML-I/II/III
Purification	Purification and separation of lectins by fast protein chromatography on Mono S cation exchanger; equilibration using 0.015 M citrate buffer, pH 4.2; elution using same buffer containing NaCl (gradient 0–0.5 M)	ML-I, ML-II, ML-III =/>95%*, Viscotoxin content <1%
*Isolation of VisalbCBA***		
	For details, see [26]	VisalbCBA >95%**, free of MLs

* Further purification by repeated affinity chromatography and fast protein chromatography. ** *Viscum album* chitin-binding agglutinin, purified by repeated affinity chromatography on a chitin carrier. VisalbCBA is able to interact with the glycan parts of ML-I, ML-II and ML-III.

against the tumor. This view is supported by a recent clinical study in melanoma patients, in which immunostimulation seemed to be the primary mode of action [24].

As in many other projects, not all important results have been published in a way such that the results of the research are easily accessible. This is particularly true for the procedure for isolation of all three MLs, which has only been published in a conference proceeding volume in India [25]. As the swan song of Uwe Pfüllers' seminal work, we therefore present the isolation procedure in tabular form so that it will be easily accessible (table 1). We also convey an even more important note to the reader: This collaboration was a lot of fun for us both for 25 years, during which a professional relationship developed into a personal friendship!

References

1 Schumacher U, Schumacher D, Schwarz T, Pfüller U: Cell biological and immunopharmacological investigations on the use of mistletoe lectin I (ML-I); in Loew D, Rietbrock N (eds): Phytopharmaka II Forschung und Klinische Anwendung. Darmstadt, Steinkopff Verlag, 1996, pp 197–204.

2 Suzuki H, Franz H, Yamamoto T, Iwasaki Y, Konno H: Identification of the normal microglial population in human and rodent nervous tissue using lectin histochemistry. Neuropathol Appl Neurobiol 1988;14:221–227.

3 Schumacher U, Kretzschmar H, Pfüller U: Staining of cerebral amyloid plaque glycoproteins in patients with Alzheimer's disease with the microglial specific lectin from mistletoe. Acta Neuropathol 1994;87:422–424.

4 Schumacher U, Adam E, Kretzschmar H, Pfüller U: Binding patterns of mistletoe lectins I, II and III to microglia and Alzheimer plaque glycoproteins in human brains. Acta Histochem 1994;96:399–403.

5 Nanka O, Peumans WJ, van Damme EJM, Pfüller U, Valasek P, Halata Z, Schumacher U, Grim M: Lectin histochemistry of microvascular endothelium in chicken and quail musculature. Anat Embryol 2001;204:407–411.

6 Jörns J, Mangold U, Neumann U, Van Damme EJM, Peumans WJ, Pfüller U, Schumacher U: Lectin histochemistry of the lymphoid organs of the chicken. Anat Embryol 2003;207:85–94.

7 Pohlmeyer I, Jörns J, Schumacher U, Van Damme EJM, Peumans WJ, Pfüller U, Neumann U: Lectin histochemical investigations of the distal gut of chicken with special emphasis on the follicle-associated epithelium. J Vet Med A 2005;52:138–146.

8 Leathem AJ, Brooks SA: Predictive value of lectin binding on breast-cancer recurrence and survival. Lancet 1987;1:1054–1056.

9 Brooks SA, Leathem AJ: Prediction of lymph node involvement in breast cancer by detection of altered glycosylation in the primary tumour. Lancet 1991;338:71–74.

10 Schumacher U, Higgs D, Loizidou M, Pickering R, Leathem A, Taylor I: Helix pomatia agglutinin binding is a useful prognostic indicator in colorectal carcinoma. Cancer 1994;74:3104–3107.

11 Dixon A, Schumacher U, Pfüller U, Taylor I: Is the binding of mistletoe lectins I and III a useful prognostic indicator in colorectal carcinoma? Eur J Surg Oncol 1994;20:648–652.

12 Thies A, Moll I, Berger J, Schumacher U: Lectin binding to cutaneous malignant melanoma: HPA is associated with metastasis formation. Br J Cancer 2001;84:819–823.

13 Thies A, Pfüller U, Schachner M, Horny H-P, Moll I, Schumacher U: Binding of mistletoe lectins to cutaneous malignant melanoma: implications for prognosis and therapy. Anticancer Res 2001;21:2883–2887.

14 Thies A, Berlin A, Brunner G, Schulze HJ, Moll I, Pfüller U, Wagener C, Schachner M, Altevogt P, Schumacher U: Glycoconjugate profiling of primary melanoma and its sentinel node and distant metastases: implications for diagnosis and pathophysiology of metastases. Cancer Lett 2007;248:68–80.

15 Thies A, Nugel D, Pfüller U, Moll I, Schumacher U: Influence of mistletoe lectins and cytokines induced by them on the cell proliferation of human melanoma cells in vitro. Toxicology 2005;207:105–116.

16 Gabius HJ, Darro F, Remmelink M, André S, Kopitz J, Danguy A, Gabius S, Salmon I, Kiss R: Evidence for stimulation of tumor proliferation in cell lines and histotypic cultures by clinically relevant low doses of the galactoside-binding mistletoe lectin, a component of proprietary extracts. Cancer Invest 2001;19:114–126.

17 Thies A, Dautel P, Meyer A, Pfüller U, Schumacher U: Low-dose mistletoe lectin-I reduces melanoma growth and spread in a scid mouse xenograft model. Br J Cancer 2008;98:106–112.

18 Schumacher U, Stamouli A, Adam E, Peddie M, Pfüller U: Biochemical, histochemical and cell biological investigations on the actions of mistletoe lectins I, II and III with human breast cancer cell lines. Glycoconj J 1995;12:250–257.

19 Freudlsperger C, Thies A, Pfüller U, Schumacher U: The proteasome inhibitor bortezomib augments anti-proliferative effects of mistletoe lectin-I and the PPAR-gamma agonist rosiglitazone in human melanoma cells. Anticancer Res 2007;27:207–213.

20 Mukthar D, Pfüller U, Tonevitsky AG, Witthohn K, Schumacher U: Cell biological investigations on the use of mistletoe lectins in cancer therapy; in Bardocz S, Pfüller U, Pusztai A (eds): COST 98 Effects of Antinutrients on the Nutritional Value of Legume Diets. Luxembourg, Office for Official Publications of the European Communities, 1998, vol V, pp 187–193.

21 Valentiner U, Pfüller U, Baum C, Schumacher U: The cytotoxic effect of mistletoe lectins I, II and III on sensitive and multidrug resistant human colon cancer cell lines in vitro. Toxicology 2002;171:187–199.

22 Müthing J, Meisen I, Bulau P, Langer M, Witthohn K, Lentzen H, Neumann U, Peter-Katalinić J: Mistletoe lectin I is a sialic acid-specific lectin with strict preference to gangliosides and glycoproteins with terminal Neu5Ac alpha 2-; 6Gal beta 1-4GlcNAc residues. Biochemistry 2004;43:2996–3007.

23 Nehmann N, Schade UM, Pfüller U, Schachner M, Schumacher U: Mistletoe lectin binds to multidrug resistance-associated protein MRP5. Anticancer Res 2009;29:4941–4948.

24 Trefzer U, Gutzmer R, Wilhelm T, Schenck F, Kähler KC, Jacobi V, Witthohn K, Lentzen H, Mohr P: Treatment of unresectable stage IV metastatic melanoma with aviscumine after anti-neoplastic treatment failure: a phase II, multi-centre study. J Immunother Cancer 2014;2:27.

25 Eifler R, Pfüller K, Göckeritz W, Pfüller U: Improved procedures for isolation and standardization of mistletoe lectins and their subunits: lectin pattern of the European mistletoe; in Chakrabarti P (ed): Lectins: Biology, Biochemistry Clinical Biochemistry. New Delhi, M/S Wiley Eastern Limited, 1994, vol 9, pp 144–151.

26 Peumans W, Verhaert P, Pfüller U, van Damme E: Isolation and partial characterization of a small chitin-binding lectin from mistletoe (Viscum album). FEBS Lett 1996;396:261–265.

Prof. Dr. Udo Schumacher
Institute of Anatomy and Experimental Morphology, University Cancer Center Hamburg
University Medical Center Hamburg-Eppendorf
Martinistraße 52, DE–20246 Hamburg (Germany)
E-Mail uschumac@uke.de

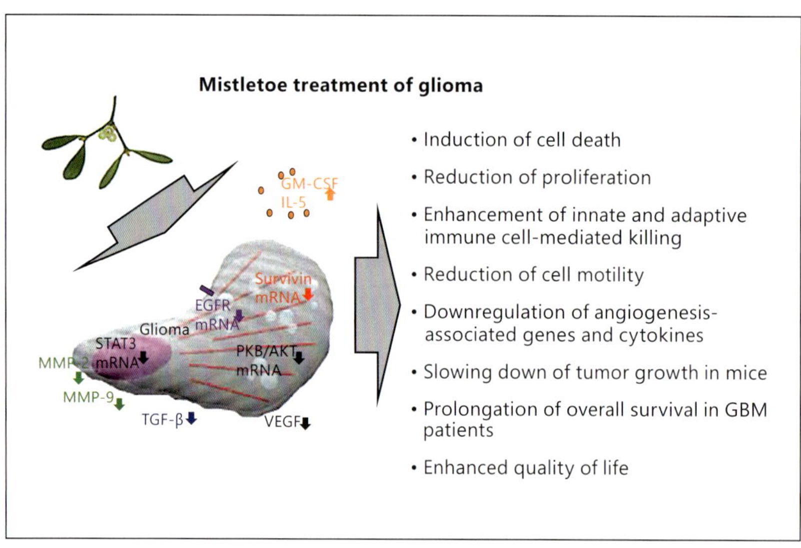

Fig. 1. Mistletoe treatment of glioma.

Mistletoe Treatment Alters Gene Expression

For glioma, it has been previously published that ISCADOR Q, a commercially available fermented extract from the European white-berry mistletoe growing on oak trees, provides multimodal anti-cancer effects both in vitro and in mouse GBM models. Growth inhibition and induction of apoptosis by ME have been demonstrated not only for GBM but also for other tumor entities, such as head and neck, breast or gynecological cancer [12–14]. Beyond direct induction of cell death and mitigation of proliferation, the expression of a variety of genes involved in GBM malignancy is altered upon mistletoe treatment. Interestingly, even though ML-I is described as a ribosome-inactivating protein, it can upregulate gene expression. In GBM cells, ISCADOR Q treatment leads to the re-expression of genes usually downregulated in this tumor but reduces the expression of genes associated with its malignant progression, pointing to a therapeutic effect of ME on gene clusters linked to gliomagenesis, thereby exerting anti-cancer effects. In GBM cells, the expression of genes regulating proliferation, such as *signal transducer and activator of transcription 3*, *protein kinase B* or *epidermal growth factor receptor*, as well as genes regulating survival (BIRC5/*survivin*) and especially secreted factors essential for cell motility, like *transforming growth factor b* (TGF-β), *tissue inhibitor of matrix metalloproteinase 2* (TIMP-2) and *matrix metalloproteinase 2* (MMP-2), is mitigated by ISCADOR Q. Additionally, the expression of multifunctional genes associated with cell adhesion, motility or angiogenesis, such as *vascular endothelial growth factor* (VEGF), *angiopoietin 1* (Ang-1) or integrins, is also reduced by ISCADOR Q (fig. 1; table 1; [12]).

Table 1. ISCADOR Q-mediated regulation of tumor-associated gene expression in GBM cells

Proliferation/survival			Immunomodulation			Neoangiogenesis			Cell motility		
protein	gene	mRNA [fold change]	protein	gene	mRNA [fold change]	protein	gene	mRNA [fold change]	protein	gene	mRNA [fold change]
HER2/neu	ERBB2	−38.1	Heme oxygenase 1	HMOX1	−33.1	Ephrin type-B receptor 4	EPHB4	−163.8	Integrin α5	ITGA5	−71.6
Protein kinase B/AKT	AKT1	−42.7	TGF-β1	TGFB1	−21.4	Integrin α5	ITGA5	−71.6	Matrix metallo-proteinase 14	MMP14	−53.3
c-Met	MET	−19.0	CD31	PECAM1	−4.4	Endoglin	ENG	−52.2	c-Met	MET	−19.0
Survivin	BIRC5	−4.7				Vascular endothelial growth factor receptor 2	VEGFR2/KDR	−37.1	Hypoxia-inducible factor 1α	HIF1A	−13.1
p53	TP53	−3.8				TGF-β1	TGFB1	−21.4	Matrix metallo-proteinase 2	MMP2	−9.9
Signal transducer and activator of transcription 3	STAT3	−3.6				c-Met	MET	−19.0	Matrix metallo-proteinase 9	MMP9	−7.0
						Thrombospondin 1	THBS1	−11.7			
						Vascular endothelial growth factor A	VEGFA	−10.9			
						Angiopoietin 1	ANGPT1	−3.7			
						Angiopoietin-like 1	ANGPTL1	−3.4			
						Angiopoietin 4	ANGPT4	−1.9			

Mistletoe Extracts as Immunomodulatory Drugs in the Treatment of Glioblastoma Multiforme

Suppression of cellular immunity is commonly seen in GBM patients, and the subsequent immunosuppression contributes to poor tumor-specific responses in affected individuals. Overexpression of TGF-β is a typical feature of GBM in vitro and in vivo. Elevated concentrations of TGF-β are detectable in glioma cyst fluids and in the cerebrospinal fluid of GBM patients, and its secretion correlates with increasing grades of glioma malignancy [15]. TGF-β plays a key role in glioma-mediated immunosuppression via blocking the activation of immune cells by downregulation of immunoactivating ligands on GBM cells and induction of apoptosis in T cells and natural killer (NK) cells. Additionally, TGF-β secretion leads to the enrichment of immunosuppressive regulatory T cells in the tumor as well as to reduced *major histocompatibility complex* expression on the tumor cells, inhibition of dendritic cell maturation and suppression of T cell proliferation [16, 17]. As mentioned before, in GBM cells, ISCADOR Q reduces TGF-β mRNA and subsequently secreted TGF-β protein levels, thus reverting TGF-β-induced immunosuppression. Additionally, ISCADOR Q tightens the interaction of GBM and NK cells, leading to enhanced NK cell-mediated GBM cell lysis [12]. Mistletoe-mediated immunostimulatory effects seem not to be specific to GBM since it has been shown that MLs provide immunomodulatory activity associated with the activation of T cells, NK cells and macrophages [18, 19] as well as with enhanced cytokine release in general [20].

Mistletoe and Cell Motility

GBM cells are particularly motile and can travel through the brain parenchyma, often following anatomical structures like vessels or axons, which guide migration. Complex interactions between glioma cells, adjacent cells and the surrounding extracellular matrix (ECM) as well as considerable changes in the cytoskeleton are prerequisites for cellular movement. The pattern of movement is governed by the ECM, which is remodeled during glioma cell invasion. The tight regulation of critical levels of proteases such as MMPs as well as of secreted factors or surface molecules on cells is necessary for the cells' physiological behavior. Changes in the balance of protein expression, such as upregulation of TGF-β or MMPs and alterations in integrin composition, determine mobile GBM cell behavior and influence the complex process of glioma cell motility [21]. Recently, it has been shown that ISCADOR Q mitigates the migration of as well as invasion by GBM, at least in vitro [12]. The multimodal effects of ME in GBM cells might be due to a couple of physiological changes induced by this treatment. Reduced secretion of MMP-2, which is one of the most important proteins for ECM destruction and a well-known inducer of GBM cell motility [22, 23], is accompanied by reduction of the activity of either pro- or mature MMP-2 as well as of MMP-9. Additionally, TIMP-2, which is able to act as an activator of MMP-2 [24], is downregulated in ISCADOR Q-treated GBM cells, which fits well with decreased MMP-2 activity and impaired cell motility [12]. The secretion of additional multifunctional factors related to GBM motility is also altered in ISCADOR Q-treated GBM cells: VEGF, whose expression in GBM is known to be correlated with MMP-2 [25]; Ang-1, a factor suggested to regulate not only angiogenesis but also GBM cell adhesion to the ECM via its receptor Tie2 [26]; and TGF-β, which, in addition to its immunosuppressive function, regulates both TIMP-2 and MMP expression and induces GBM cell migration and invasion [27]. Reduced expression of TGF-β, VEGF and Ang-1 might therefore be suggested to contribute not only to the anti-migratory effect of ISCADOR Q but also to reduced tumor vascularization and neoangiogenesis. Even the anti-migratory effects of ME seem not to be specific to GBM since it was recently reported that ME treatment also inhibits the migration of breast cancer cells [14].

Therapeutic Effects of Mistletoe in Experimental Rodent Glioblastoma Multiforme Models

To evaluate the therapeutic effects of mistletoe components that have been recorded in many in vitro studies, rodent glioma models are feasible for analyzing antitumoral effects in vivo. So far, mouse and rat xenograft models as well as syngeneic rodent models have been used for mistletoe treatment of either subcutaneous or orthotopically grown GBM. In subcutaneous xenografts as well as in syngeneic GBM-bearing mice, either systemic or intratumoral application of ISCADOR Q reduced tumor growth, with the effects of intratumoral ISCADOR Q injection being significantly superior to its systemic application [12]. Similar results were obtained in a rat glioma model using systemic or intratumoral injection of purified galactoside-specific ML [28]. So far, the anti-cancer activities of mistletoe compounds have been tested in rodents in a variety of other tumor entities. In a mouse leukemia model, mistletoe treatment improved overall survival. In mouse melanoma xenografts, ML-I reduced the tumor mass and metastasis and led to an increased infiltration of dendritic cells into the tumor, and in murine models of breast, pancreatic, bladder, lung and liver cancer, ME treatment reduced tumor growth [29–33].

As it is known that the half-life of ML-I is very short [34] and that ML-I, in addition to being immunomodulatory, has direct cytotoxicity, systemic treatment with mistletoe-based drugs might be not the optimal route of application. Direct contact of mistletoe compounds with tumor cells, as happens following intratumoral injection or infusion, should be considered as an alternative.

Clinical Trials Using Mistletoe for the Treatment of Glioma Patients

To date, two clinical trials using mistletoe to treat patients with World Health Organization grade III–IV glioma have been published. All patients underwent surgical resection as well as radiochemotherapy prior to ML-I treatment. One of these trials demonstrated strong immunomodulatory effects after subcutaneous injection of ML-I. Additionally, patients in the mistletoe treatment group reported an improvement in their quality of life [35, 36]. This result seems not to be specific to GBM patients since during mistletoe treatment, cancer patients harboring other types of cancer reported similar effects as well as reduced side effects for the conventional therapy, like fatigue, vomiting and nausea. Additionally, these patients reported being less depressive and feeling better emotionally and were endued with higher levels of energy and an enhanced ability to work. However, the mechanism responsible for these positive psychological effects of mistletoe treatment remains unknown [36]. A second clinical trial using systemic ML-I treatment in glioma patients demonstrated a tendency toward prolonged relapse-free survival and a significant prolongation of overall survival in the treatment group [37]. Unfortunately, both clinical trials did not distinguish between World Health Organization grades III and IV or between primary (de novo) and recurrent glioma, were not blinded and did not contain a control placebo treatment group. Due to the diversity of mistletoe products (e.g. extract vs. single compound) and their varying proportion of pharmacologically relevant constituents, even interpretation of the clinical studies is difficult. Nonetheless, a pooled analysis of clinical studies using mistletoe-based drugs as cancer therapeutics revealed better survival among cancer patients after mistletoe treatment [38]. Keeping in mind all of these limitations, future studies evaluating the effects of mistletoe compounds on the treatment of glioma should focus on well-designed treatment protocols to obtain complete and reliable data sets.

Conclusion

In recent years, an increasing number of publications showing data from in vitro studies, rodent cancer models and clinical trials have clearly pointed out that mistletoe-based drugs are agents providing anti-cancer activity if used as concomitant therapeutics in parallel to standard therapy. Nevertheless, future studies evaluating the effects of mistletoe compounds in the treatment of cancer should focus on more evidence-based preclinical data, higher quality, a transparent study design and clear endpoints to provide greater insight into a treatment that is often deprecated as not effective. The information from these future studies should be taken seriously to provide optimal care to cancer patients.

References

1 Stupp R, Mason WP, van den Bent MJ, Weller M, Fisher B, Taphoorn MJ, Belanger K, Brandes AA, Marosi C, Bogdahn U, Curschmann J, Janzer RC, Ludwin SK, Gorlia T, Allgeier A, Lacombe D, Cairncross JG, Eisenhauer E, Mirimanoff RO: Radiotherapy plus concomitant and adjuvant temozolomide for glioblastoma. N Engl J Med 2005;352:987–996.

2 Vartanian A, Singh SK, Agnihotri S, Jalali S, Burrell K, Aldape KD, Zadeh G: GBM's multifaceted landscape: highlighting regional and microenvironmental heterogeneity. Neuro Oncol 2014;16:1167–1175.

3 Avril T, Vauleon E, Tanguy-Royer S, Mosser J, Quillien V: Mechanisms of immunomodulation in human glioblastoma. Immunotherapy 2011;3:42–44.

4 Hajto T, Hostanska K, Saller R: Mistletoe therapy from the pharmacologic perspective. Forsch Komplementarmed 1999;6:186–194.

5 Krauspenhaar R, Eschenburg S, Perbandt M, Kornilov V, Konareva N, Mikailova I, Stoeva S, Wacker R, Maier T, Singh T, Mikhailov A, Voelter W, Betzel C: Crystal structure of mistletoe lectin I from Viscum album. Biochem Biophys Res Commun 1999;257: 418–424.

6 Puri M, Kaur I, Perugini MA, Gupta RC: Ribosome-inactivating proteins: current status and biomedical applications. Drug Discov Today 2012;17:774–783.

7 Elluru S, Duong Van Huyen JP, Delignat S, Prost F, Bayry J, Kazatchkine MD, Kaveri SV: Molecular mechanisms underlying the immunomodulatory effects of mistletoe (Viscum album L.) extracts Iscador. Arzneimittelforschung 2006;56:461–466.

8 Lee CH, Kim JK, Kim HY, Park SM, Lee SM: Immunomodulating effects of Korean mistletoe lectin in vitro and in vivo. Int Immunopharmacol 2009;9: 1555–1561.

9 Stein GM, Schaller G, Pfuller U, Schietzel M, Büssing A: Thionins from Viscum album L: influence of the viscotoxins on the activation of granulocytes. Anticancer Res 1999;19:1037–1042.

10 Giudici AM, Regente MC, Villalain J, Pfuller K, Pfuller U, De La Canal L: Mistletoe viscotoxins induce membrane permeabilization and spore death in phytopathogenic fungi. Physiol Plant 2004;121:2–7.

11 Stein GM, Schaller G, Pfuller U, Wagner M, Wagner B, Schietzel M, Büssing A: Characterisation of granulocyte stimulation by thionins from European mistletoe and from wheat. Biochim Biophys Acta 1999; 1426:80–90.

12 Podlech O, Harter PN, Mittelbronn M, Pöschel S, Naumann U: Fermented mistletoe extract as a multimodal antitumoral agent in gliomas. Evid Based Complement Alternat Med 2012;2012:501796.

13 Klingbeil MF, Xavier FC, Sardinha LR, Severino P, Mathor MB, Rodrigues RV, Pinto DS Jr: Cytotoxic effects of mistletoe (Viscum album L.) in head and neck squamous cell carcinoma cell lines. Oncol Rep 2013;30:2316–2322.

14 Kienle GS, Glockmann A, Schink M, Kiene H: Viscum album L. extracts in breast and gynaecological cancers: a systematic review of clinical and preclinical research. J Exp Clin Cancer Res 2009;28:79.

15 Kjellman C, Olofsson SP, Hansson O, Von Schantz T, Lindvall M, Nilsson I, Salford LG, Sjogren HO, Widegren B: Expression of TGF-beta isoforms, TGF-beta receptors, and SMAD molecules at different stages of human glioma. Int J Cancer 2000;89: 251–258.

16 Platten M, Wick W, Weller M: Malignant glioma biology: role for TGF-beta in growth, motility, angiogenesis, and immune escape. Microsc Res Tech 2001; 52:401–410.

17 Eisele G, Wischhusen J, Mittelbronn M, Meyermann R, Waldhauer I, Steinle A, Weller M, Friese MA: TGF-beta and metalloproteinases differentially suppress NKG2D ligand surface expression on malignant glioma cells. Brain 2006;129:2416–2425.

18 Yoon TJ, Yoo YC, Kang TB, Song SK, Lee KB, Her E, Song KS, Kim JB: Antitumor activity of the Korean mistletoe lectin is attributed to activation of macrophages and NK cells. Arch Pharm Res 2003;26:861–867.

19 Fischer S, Scheffler A, Kabelitz D: Stimulation of the specific immune system by mistletoe extracts. Anticancer Drugs 1997;8:S33–S37.

20 Huber R, Rostock M, Goedl R, Ludtke R, Urech K, Buck S, Klein R: Mistletoe treatment induces GM-CSF- and IL-5 production by PBMC and increases blood granulocyte- and eosinophil counts: a placebo controlled randomized study in healthy subjects. Eur J Med Res 2005;10:411–418.

21 Goldbrunner RH, Bernstein JJ, Tonn JC: Cell-extracellular matrix interaction in glioma invasion. Acta Neurochir 1999;141:295–305.

22 Deryugina EI, Bourdon MA, Luo GX, Reisfeld RA, Strongin A: Matrix metalloproteinase-2 activation modulates glioma cell migration. J Cell Sci 1997;110: 2473–2482.

23 Guo P, Imanishi Y, Cackowski FC, Jarzynka MJ, Tao HQ, Nishikawa R, Hirose T, Hu B, Cheng SY: Upregulation of angiopoietin-2, matrix metalloprotease-2, membrane type 1 metalloprotease, and laminin 5 gamma 2 correlates with the invasiveness of human glioma. Am J Pathol 2005;166:877–890.

24 Lu KV, Jong KA, Rajasekaran AK, Cloughesy TF, Mischel PS: Upregulation of tissue inhibitor of metalloproteinases (TIMP)-2 promotes matrix metalloproteinase (MMP)-2 activation and cell invasion in a human glioblastoma cell line. Lab Invest 2004;84:8–20.

25 Munaut C, Noel A, Hougrand O, Foidart JM, Boniver J, Deprez M: Vascular endothelial growth factor expression correlates with matrix metalloproteinases MT1-MMP, MMP-2 and MMP-9 in human glioblastomas. Int J Cancer 2003;106:848–855.

26 Lee OH, Xu J, Fueyo J, Fuller GN, Aldape KD, Alonso MM, Piao Y, Liu TJ, Lang FF, Bekele BN, Gomez-Manzano C: Expression of the receptor tyrosine kinase Tie2 in neoplastic glial cells is associated with integrin beta1-dependent adhesion to the extracellular matrix. Mol Cancer Res 2006;4:915–926.

27 Wick W, Naumann U, Weller M: Transforming growth factor-beta: a molecular target for the future therapy of glioblastoma. Curr Pharm Des 2006;12:341–349.

28 Lenartz D, Andermahr J, Plum G, Menzel J, Beuth J: Efficiency of treatment with galactoside-specific lectin from mistletoe against rat glioma. Anticancer Res 1998;18:1011–1014.

29 Beuth J, Ko HL, Schneider H, Tawadros S, Kasper HU, Zimst H, Schierholz JM: Intratumoral application of standardized mistletoe extracts down regulates tumor weight via decreased cell proliferation, increased apoptosis and necrosis in a murine model. Anticancer Res 2006;26:4451–4456.

30 Seifert G, Jesse P, Laengler A, Reindl T, Luth M, Lobitz S, Henze G, Prokop A, Lode HN: Molecular mechanisms of mistletoe plant extract-induced apoptosis in acute lymphoblastic leukemia in vivo and in vitro. Cancer Lett 2008;264:218–228.

31 Thies A, Dautel P, Meyer A, Pfuller U, Schumacher U: Low-dose mistletoe lectin-I reduces melanoma growth and spread in a scid mouse xenograft model. Br J Cancer 2008;98:106–112.

32 Rostock M, Huber R, Greiner T, Fritz P, Scheer R, Schueler J, Fiebig HH: Anticancer activity of a lectin-rich mistletoe extract injected intratumorally into human pancreatic cancer xenografts. Anticancer Res 2005;25:1969–1975.

33 Mengs U, Schwarz T, Bulitta M, Weber K: Antitumoral effects of an intravesically applied aqueous mistletoe extract on urinary bladder carcinoma MB49 in mice. Anticancer Res 2000;20:3565–3568.

34 Schöffski P, Riggert S, Fumoleau P, Campone M, Bolte O, Marreaud S, Lacombe D, Baron B, Herold M, Zwierzina H, Wilhelm-Ogunbiyi K, Lentzen H, Twelves C: Phase I trial of intravenous aviscumine (rViscumin) in patients with solid tumors: a study of the European Organization for Research and Treatment of Cancer New Drug Development Group. Ann Oncol 2004;15:1816–1824.

35 Lenartz D, Stoffel B, Menzel J, Beuth J: Immunoprotective activity of the galactoside-specific lectin from mistletoe after tumor destructive therapy in glioma patients. Anticancer Res 1996;16:3799–3802.

36 Kienle GS, Kiene H: Review article: influence of Viscum album L (European mistletoe) extracts on quality of life in cancer patients: a systematic review of controlled clinical studies. Integr Cancer Ther 2010;9:142–157.

37 Lenartz D, Dott U, Menzel J, Schierholz JM, Beuth J: Survival of glioma patients after complementary treatment with galactoside-specific lectin from mistletoe. Anticancer Res 2000;20:2073–2076.

38 Ostermann T, Raak C, Büssing A: Survival of cancer patients treated with mistletoe extract (Iscador): a systematic literature review. BMC Cancer 2009;9:451.

Ulrike Naumann
Hertie Institute for Clinical Brain Research, Department of Vascular Neurology
Molecular Neuro-Oncology, University of Tübingen
Otfried-Müller-Str. 27, DE–72076 Tübingen (Germany)
E-Mail ulrike.naumann@uni-tuebingen.de

Zänker KS, Kaveri SV (eds): Mistletoe: From Mythology to Evidence-Based Medicine.
Transl Res Biomed. Basel, Karger, 2015, vol 4, pp 57–66 (DOI: 10.1159/000375426)

Cancer Surgery and Supportive Mistletoe Therapy: From Scepticism to Randomised Clinical Trials

Danijel Galun[a, b] · Wilfried Tröger[c] · Miroslav Milicevic[a, b]

[a]Clinic for Digestive Surgery, Clinical Center of Serbia, [b]Medical School, University of Belgrade, Belgrade, Serbia; [c]Clinical Research Dr. Tröger, Freiburg, Germany

Abstract

Surgery remains the core of curative-intent cancer treatment in a world of multidisciplinary therapies using different treatment modalities. Despite innovative approaches to cancer destruction, including surgery, chemotherapy, radiotherapy and hormone therapy, cancer mortality rates have not been significantly reduced in the US or other developed countries in the past decades. While conventional oncology has introduced therapeutic innovations, complementary approaches have also been examined in clinical trials. Mistletoe extract therapy is among the most thoroughly studied complementary treatments in Europe. Several systematic reviews and meta-analyses have found that mistletoe treatment is beneficial to cancer patients in terms of survival, improved quality of life and minimised side effects of anticancer chemotherapy. Several studies have examined the effects of mistletoe treatment in different perioperative settings, demonstrating the positive effects of mistletoe extract therapy. This chapter focuses on mistletoe treatment in pancreatic cancer patients, beginning with case reports and progressing to randomised clinical trials, finally demonstrating that dedicated clinical practice and evidence obtained from studies of appropriate methodological quality can turn scepticism into objective treatment for the benefit of cancer patients.

© 2015 S. Karger AG, Basel

Introduction

Although modern cancer treatment is based on a multidisciplinary approach using different treatment modalities, surgery remains the core of curative-intent cancer treatment. Whether operating on patients with resectable malignant disease or deciding on palliative treatment for patients with nonresectable disease, surgeons notably remain on the front lines in the struggle against cancer. However, despite innovative approaches to cancer destruction, including surgery, chemotherapy, radiotherapy

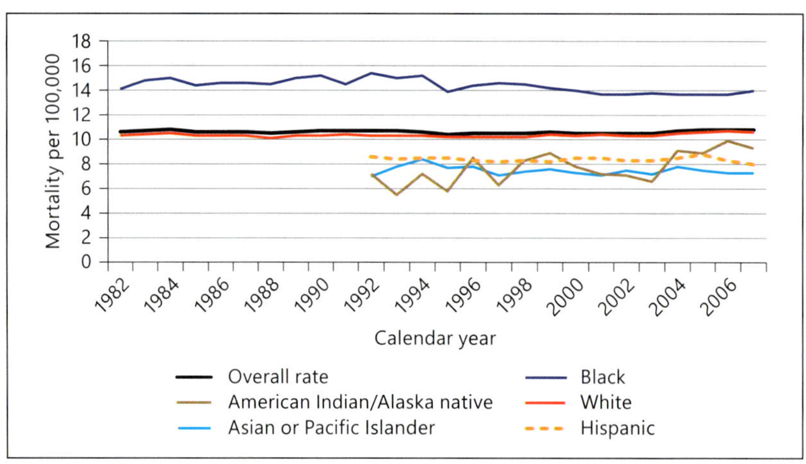

Fig. 1. Mortality from pancreatic cancer in the US from 1982 to 2007 [7].

and hormone therapy, cancer mortality rates have not significantly decreased in the US or other developed countries in the past decades [1–4].

Several authors have dampened the optimism associated with chemotherapy for advanced carcinoma, especially when responses are used as measures of therapeutic success, and they have urged the medical community to think about new therapeutic strategies [1, 2].

Some malignant diseases, like pancreatic and hepatocellular cancers, are particularly resistant to any treatment modality. Still, in the 21st century, the number of new pancreatic and hepatocellular cancer patients diagnosed annually is equal to the number of deaths from these two cancers, indicating that national health systems and the worldwide community have no efficient solution to this problem [3–7]. The US National Cancer Institute report is particularly depressing, showing that the marked increase in the funding of pancreatic cancer research, from 21.8 to 89.7 million USD, from 2001 to 2009 as well as the increases in the number of active projects (319 in 2009) and in the number of scientific publications on pancreatic cancer (422 in 2009) have had no impact on mortality from pancreatic cancer, which has remained unchanged for years [7] (fig. 1).

Conventional Oncology and Mistletoe

While conventional oncology has introduced therapeutic innovations like targeted antibody therapies, specific inhibitors of cancer cell growth factors, new chemotherapeutic agents and advanced radiotherapy modalities, other nontoxic, complementary approaches have also been examined in clinical trials [8], and surgeons should consider different approaches.

According to several authors, plant lectins, i.e. carbohydrate-binding proteins, show remarkable antitumor properties [9–11]. Their potential as antineoplastic drugs needs to be thoroughly examined from the bench to the clinic [9].

Mistletoe extract therapy is among the most thoroughly studied complementary treatments in Europe [8]. Several systematic reviews and meta-analyses have found a benefit from mistletoe treatment on cancer patient survival [12, 13], improving quality of life [12, 14, 15] and minimising the side effects of anticancer chemotherapy [14].

Other challenging studies have examined the effects of mistletoe treatment in different perioperative settings. In a randomised clinical trial, Schink et al. demonstrated that perioperative infusion of Iscador® (Weleda, Switzerland) can prevent the suppression of natural killer cell activity associated with major surgery, which potentially may favour haematogenic tumour cell dissemination [16]. The impact of this therapy on the relapse and survival of cancer patients should be tested in additional clinical trials.

Quality of Life and Mistletoe

Patient quality of life and the safety of adjuvant mistletoe treatment were tested in gastric cancer patients in a randomised clinical trial [17]. Thirty-two operated gastric cancer patients (stage Ib or II) who were scheduled for oral chemotherapy were randomised 1:1 to receive additional mistletoe extract therapy or no additional therapy. The additional mistletoe treatment was found to be safe and was associated with improved quality of life in these gastric cancer patients [17]. Seventy patients with different digestive tract cancers were included in a study that examined the perioperative use of the mistletoe drug Isorel® (Novipharm, Austria). It was demonstrated that the additional mistletoe therapy was associated with improved immune competence and an improved overall health status of cancer patients undergoing surgery [18].

Cochrane Reviews and Mistletoe

However, Cochrane reviewers found that out of 80 mistletoe studies examined for the purpose of assessing mistletoe therapy in oncology, 58 had no prospective trial design with randomised treatment allocation, and these were excluded from the analysis [19]. Among 13 trials investigating survival, 6 showed some evidence of a benefit, but none of them was of high methodological quality. Among 16 trials investigating the efficacy of mistletoe extracts in improving quality of life, psychological parameters, the performance index, symptom scales or the adverse effects of chemotherapy, only 2 were of superior methodological quality. Although 14 previously published studies providing grade I and II evidence on survival, tumour behaviour and quality of life were not included in the review, the overall conclusion remained that more high-quality, independent clinical research is needed to truly assess the safety and effectiveness of mistletoe extracts [19].

Following the overall conclusion from the Cochrane review, a randomised clinical trial was conducted in a high-volume surgical unit among patients with advanced or metastatic pancreatic cancer (mPC) to assess the safety and efficacy of mistletoe extract therapy using the mistletoe drug Iscador® Qu [20]. This study demonstrated that dedicated clinical practice and evidence obtained from studies with appropriate methodological quality could turn scepticism into objective treatment for the benefit of cancer patients. Here, we present a short review of mistletoe treatment in pancreatic cancer patients.

Pancreatic Cancer and Mistletoe: From Single Cases to Randomised Clinical Trials

A search was performed on July 15, 2014, using specific key words (*Viscum album* or mistletoe or Iscador or Helixor or Abnoba or Iscucin or Vysorel or Isorel or Eurixor or Lektinol or Plenosol and pancreas or pancreatic or pancreas) in the following databases: Adis Newsletters (AN83); Deutsches Ärzteblatt (AR96); GLOBAL Health (AZ72); BIOSIS Previews (BA00); BIOSIS Previews (BA26); AMED (CB85); CCMED (CC00); Cochrane Central Register of Controlled Trials (CCTR93); Cochrane Database of Systematic Reviews (CDSR93); CAB Abstracts (CV72); Derwent Drug File (DD83); EMBASE Alert (EA08); EMBASE (EM00); EMBASE (EM47); IPA (IA70); ISTPB + ISTP/ISSHP (II01); ISTPB + ISTP/ISSHP (II78); Health Technology Assessment Database (INAHTA); SciSearch (IS00); SciSearch (IS74); MEDLINE (ME00); MEDLINE (ME60); and NLM GATEWAY/PUBMED/MEDLINE, databases of manufacturers of mistletoe extracts and private databases. This search yielded 272 hits, of which 141 were unique, and 94 were related to pancreatic cancer. In addition to 11 publications of laboratory results, 22 comments and 7 reviews, 54 clinical reports were found. Among them, 5 case reports, 26 abstracts or posters and 23 study reports were identified. Because multiple publications were identified, we have cited only 23 of the 54 publications.

The first case report was published in 1967 and described a patient with inoperable mPC. Besides Iscador®, the patient received Formica D4-D6, Stibium D6 and Chelidonium comp (Weleda, Switzerland) [21] (table 1). The patient was still alive at the time of the report, which was 7 years after diagnosis. Eight years later, a second case report on the application of Iscador® in a patient with mPC described a surprisingly long survival of 4½ years without pain [22]. Concomitant therapies were Ferrum carb. D8 per os, Cardiodoron, Chelidonium comp, Oxalis D2, Stannum per Geh. 0.1%, Ferrum carb. D6, Vitis comp, Renodoron and daily wet compresses of yarrow on the abdomen. Because of the extraordinary clinical courses of the two pancreatic adenocarcinoma patients treated with mistletoe extracts, the first retrospective study was published in 1979 [23].

Table 1. Publications of results of mistletoe therapy for pancreatic cancer sorted by publication year

Author, year [citation]	Design (objective)	Mistletoe preparation (application form)	N	Median OS, months	QoL	Disease-related symptoms	Quality measures
Leroi, 1967 [21]	Case report (survival)	Iscador (SC)	1	84	↑	↓	G
Reinitzer, 1975 [22]	Case report (survival)	Iscador (SC)	1	54	↑	↓	G
Delius-Müller, 1979 [23]	Retrospective (survival)	Iscador (SC)	80	5.9	↑	↓	G
		control	Lit	1.5	n.d.	n.d.	
Boie et al., 1981 [25]	Retrospective (survival)	Helixor (SC)	17	5.5	n.d.	n.d.	C, G, H
		control	54 Lit	1.4	n.d.	n.d.	
Buchner, 1984 [27]	Retrospective	Iscador (SC)	17	4.2	n.d.	↓	C, G, H
		control	9	4.1	n.d.	n.d.	
Hager et al., 1995 [29]	Retrospective	Eurixor, Helixor (SC)	28	12	↑	↓	C, G, H
Friess et al., 1996 [30]	Prospective (survival, QoL)	Eurixor (SC)	16	5.6	↓	n.d.	A, C, F, G, H
Schaefermeyer and Schaefermeyer, 1998 [31]	Retrospective (survival)	Iscador (SC)	320	6.8	n.d.	n.d.	C, G, H
		control	Lit	2.8–4.0	n.d.	n.d.	
Huber et al., 2000 [33]	Case report (eosinophils, QoL)	Abnobaviscum (IT)	1	n.d.	→	→	G, H
Matthes, 2001 [34]	Case series (re-occlusion)	Abnobaviscum, Helixor (IT)	3	2	n.d.	n.d.	
Enesel et al., 2005 [18]	RCT (immunology, QoL)	Isorel (SC)	2 of 40	n.d.	↑	n.d.	A, B, C, D, F, G, H
		control	3 of 30	n.d.	↓	n.d.	
Hager et al., 2009 [36]	Prospective (survival, QoL)	Eurixor, Helixor (SC)	46	10.8	↑	n.d.	C, G, H
Stumpf et al., 2009 [35]	Retrospective (survival)	Iscador, Helixor (SC)	69	9.1	n.d.	n.d.	D, H
		control	Lit.	3.2	n.d.	n.d.	
Matthes et al., 2010 [38]	Retrospective (survival)	Iscador (SC)	201 DB	15.4	↑	↓	C, D, F, H, I
		control	195 DB	11.3	–	–	
Ritter et al., 2010 [37]	Case report (survival)	mistletoe (SC)	1	18.9	n.d.	n.d.	G, H
Brandenberger et al., 2012 [40]	Prospective (QoL)	Iscador, Abnobaviscum, Isorel (SC)	1 of 3	n.d.	↑	n.d.	A, D, F, H
Wiebelitz and Beer, 2012 [41]	Case report (QoL)	Abnobaviscum, Cefalektin, Iscador (SC)	1	n.d.	↑	↓	G, H
Mansky et al., 2013 [42]	Prospective (safety)	Helixor (SC)	8	n.d.	–	–	A, C, D, E, F, G, H, I
Schad et al., 2013 [43]	Retrospective (survival)	Helixor (IT)	39	11	n.d.	n.d.	C, D, H
Steele et al., 2014 [44, 45]	Retrospective (safety)	Abnobaviscum, Helixor, Iscador, Iscucin, Isorel (SC, IV)	233	n.d.	n.d.	n.d.	D, F, H
Tröger et al., 2013, 2014 [20, 46]	RCT (survival, QoL)	Iscador (SC)	110	4.8	↑	↓	A, B, C, D, E, F, G, H, I

DB = Out of an existing database at the same centres; IT = intratumoural, IV = intravenous; Lit = out of cited literature; OS = overall survival; n.d. = not determined; – = not assessable; QoL = quality of life; RCT = prospective, randomised, controlled clinical trial; and SC = subcutaneous.
(A) Protection against performance (treatment) bias by standardisation of care protocol and documentation of all co-interventions. (B) Protection against selection bias, especially by adequate randomisation. (C) Minimisation of heterogeneity by pre-stratification or matching. (D) Protection against measurement (detection) bias by standardisation of outcome assessment. (E) Protection against attrition (exclusion) bias (lost patients <10%) or by intention-to-treat analysis (including non-adherers as randomised) plus per-protocol analysis (excluding non-adherers) in combination with sensitivity analysis and by comparison of prognostic characteristics of lost patients and compliers. (F) Well-described study design. (G) Well-described intervention, patient characteristics, and disease (diagnosis, stage, and duration). (H) Effect measurement was relevant and well described. (I) Data quality assured by the International Conference on Harmonization Good Clinical Practice guidelines, especially by monitoring.

In a retrospective study of 223 intratumoral applications of mistletoe extracts during conventional therapy in 39 patients with mPC, the median survival time was 11 months. Due to the multimodal therapeutic setting, the effect of the mistletoe treatment was unclear; however, an improvement in quality of life was seen. No severe adverse drug reactions (ADRs) were recorded, and the application was safe [43].

In a retrospective safety analysis of 1,923 cancer patients treated with subcutaneous mistletoe extracts, 233 patients suffered from pancreatic cancer. Although no stratification by tumour entity was performed, the following results determined for all of the patients may be valid for pancreatic cancer patients as well: of 1,923 patients, 283 (14.7%) reported 427 expected effects (local reactions <5 cm and an increase in body temperature <38°C). ADRs were documented in 162 (8.4%) patients, who reported a total of 264 events. The ADRs were mild (50.8%), moderate (45.1%), or severe (4.2%), and none was serious [44]. Similar findings were reported in a comparable safety analysis, as follows: of 475 cancer patients who received intravenous infusions of Helixor®, Abnobaviscum®, or Iscador® mistletoe preparations, 22 patients (4.6%) reported 32 ADRs of mild (59.4%) or moderate (40.6%) severity. Six of the 22 patients had cancer of the pancreas. Again, no serious ADRs occurred. Both safety analyses may comprise data from the patients described above [45].

Finally, 220 patients with inoperable mPC were included in a prospective, randomised clinical trial conducted according to the International Conference on Harmonization Good Clinical Practice guidelines. Two audits and one inspection of the national competent authority took place. A comparison of 110 patients receiving Iscador® with 110 patients receiving no antineoplastic therapy yielded a survival of 4.8 months (Iscador group) versus 2.7 months (control group) (hazard ratio = 0.55; p = 0.0031). Fewer disease-related symptoms of all grades and no therapy-related adverse events were observed in the Iscador group. Additionally, the patients in the Iscador group showed clinically relevant and significantly better quality of life (based on the Organization for the Research and Treatment of Cancer Quality of Life Questionnaire QLQ-C30) for 13 of 15 items, and they did not lose weight [20, 46] (table 1).

Conclusion

For patients with advanced pancreatic cancer or mPC treated with mistletoe extracts, we detailed the course of documentation, from single case reports to retrospective analyses and, finally, to randomised, controlled prospective studies. The last presented trial seems to verify all reports of lower evidence and quality, which have demonstrated longer survival and better overall quality of life, with less pain and a greater appetite associated with mistletoe treatment and with no severe therapy-related ADRs. Additional studies are needed to assess the benefits of combining different chemotherapy protocols with mistletoe preparations for pancreatic cancer patients.

References

1 Bailar JC, Gornik HL: Cancer undefeated. N Engl J Med 1997;336:1569–1574.

2 Faguet GB: The War on Cancer: An Anatomy of Failure, A Blueprint for the Future. Dordrecht, Springer, 2005.

3 National Cancer Institute: Cancer trends progress report – 2009/2010 update. September 2010. http://progressreport.cancer.gov (accessed August 29, 2014).

4 World Health Organization: Mortality database. WHO statistical information system. http://www.who.int/whosis (accessed August 29, 2014).

5 International Agency for Research on Cancer (IARC): GLOBOCAN 2012. Estimated Cancer Incidence, Mortality and Prevalence Worldwide in 2012. http://www.globocan.iarc.fr (accessed August 29, 2014).

6 Kew MC: Hepatocellular carcinoma: epidemiology and risk factors. Journal of Hepatocellular Carcinoma 2014;1:115–125.

7 National Cancer Institute: Pancreatic Cancer: A Summary of NCI's Portfolio and Highlights of Recent Research Progress. US Department of Health and Human Services, 2010.

8 Beuth J, Schierholz JM: Evidence-based complementary oncology. Innovative approaches to optimize standard therapy strategies. In Vivo 2007;21:423–428.

9 Liu B, Bian H, Bao J: Plant lectins: potential antineoplastic drugs from bench to clinic. Cancer Lett 2010;287:1–12.

10 Liu Z, Luo Y, Zhou TT, Zhang WZ: Could plant lectins become promising anti-tumour drugs for causing autophagic cell death? Cell Prolif 2013;46:509–515.

11 Zhang X, Chen LX, Ouyang L, Cheng Y, Liu B: Plant natural compounds: targeting pathways of autophagy as anti-cancer therapeutic agents. Cell Prolif 2012;45:466–476.

12 Kienle GS, Berrino F, Büssing A, Portalupi E, Rosenzweig S, Kiene H: Mistletoe in cancer – a systematic review on controlled clinical trials. Eur J Med Res 2003;8:109–119.

13 Ostermann T, Raak C, Büssing A: Survival of cancer patients treated with mistletoe extract (Iscador): a systematic literature review. BMC Cancer 2009;9:451.

14 Kienle GS, Kiene H: Influence of Viscum album L (European mistletoe) extracts on quality of life in cancer patients: a systematic review of controlled clinical studies. Integr Cancer Ther 2010;9:142–157.

15 Melzer J, Iten F, Hostanska K, Saller R: Efficacy and safety of mistletoe preparations (Viscum album) for patients with cancer diseases. A systematic review. Forschende Komplementärmedizin 2009;16:217–226.

16 Schink M, Tröger W, Dabidian A, Goyert A, Scheuerecker H, Meyer J, Fischer IU, Glaser F: Mistletoe extract reduces the surgical suppression of natural killer cell activity in cancer patients. A randomized phase III trial. Forsch Komplementärmed Klass Naturheilkd 2007;14:9–17.

17 Kim KC, Yook JH, Eisenbraun J, Kim BS, Huber R: Quality of life, immunomodulation and safety of adjuvant mistletoe treatment in patients with gastric carcinoma – a randomized, controlled pilot study. BMC Complement Altern Med 2012;12:172.

18 Enesel MB, Acalovschi I, Grosu V, Sbarcea A, Rusu C, Dobre A, Weiss T, Zarkovic N: Perioperative application of the Viscum album extract Isorel in digestive tract cancer patients. Anticancer Res 2005;25:4583–4590.

19 Horneber MA, Bueschel G, Huber R, Linde K, Rostock M: Mistletoe therapy in oncology. Cochrane Database Syst Rev 2008;2:CD003297.

20 Tröger W, Galun D, Reif M, Schumann A, Stankovic N, Milicevic M: Viscum album [L.] extract therapy in patients with locally advanced or metastatic pancreatic cancer: a randomised clinical trial on overall survival. Eur J Cancer 2013;49:3788–3797.

21 Leroi A: Fortschritte in der Iscador-behandlung maligner tumoren. Beiträge zu einer Erweiterung der Heilkunst nach Geisteswissenschaftlichen Erkenntnissen 1967;3:86–107.

22 Reinitzer HR: Mistel-studie krankengeschichten. Mitteilungen aus der Behandlung Maligner Tumoren mit Viscum Album 1975;7:26–28.

23 Delius-Müller U: Ergebnisse der Iscadortherapie bei der behandlung des pankreaskarzinoms in der Lukas-Klinik. Mitteilungen aus der Behandlung Maligner Tumoren mit Viscum Album 1979;11:5–8.

24 Heinkel K: Das pankreaskarzinom. Mitteilungen aus der Behandlung Maligner Tumoren mit Viscum Album 1979;2:2–3.

25 Boie D, Gutsch J, Burkhardt R: Die behandlung von lebermetastasen verschiedener primärtumoren mit Helixor. Therapiewoche 1981;31:1865–1869.

26 Jaffe BM, Donegan WL, Watson F, Spratt JS: Factors influencing survival in patients with untreated hepatic metastases. Surg Gynecol Obstet 1968;127:1–11.

27 Buchner C: Über den Verbrauch von Analgetika und Sedativa bei Malignompatienten mit und ohne adjuvante Therapie mit Iscador[R]; dissertation, Universität Marburg, Marburg, 1984, p 52.

28 Gudjonsson B, Livstone EM, Spiro HM: Cancer of the pancreas: diagnostic accuracy and survival statistics. Cancer 1978;42:2494–2506.

29 Hager ED, Kleef R, Popa C, Dziambor H, Schrittwieser G, Strama H, Süße B: Komplementärtherapie des nicht kurativ resektablen Pankreaskarzinoms. Dtsch Zschr Onkol 1995;27:115–123.

30 Friess H, Beger HG, Kunz J, Funk N, Schilling M, Buchler MW: Treatment of advanced pancreatic cancer with mistletoe: results of a pilot trial. Anticancer Res 1996;16:915–920.

31 Schaefermeyer G, Schaefermeyer H: Treatment of pancreatic cancer with *Viscum album* (Iscador): a retrospective study of 292 patients 1986–1996. Complement Ther Med 1998;6:172–177.

32 Pollard HM, Anderson W, Brooks FP: Staging of cancer of the pancreas. Cancer 1980;47:1631–1637.

33 Huber R, Barth H, Schmitt-Gräff A, Klein R: Hypereosinophilia induced by high-dose intratumoral and peritumoral mistletoe application to a patient with pancreatic carcinoma. J Altern Complement Med 2000;6:305–310.

34 Matthes H: Onkologische misteltherapie (*Viscum album* L.) aus klinisch-anthroposophischer sicht; in: Scheer R, Bauer R, Becker H, Berg PA, Fintelmann V (eds):Die Mistel in der Tumortherapie: Grundlagenforschung und Klinik. Essen, KVC Verlag, 2001, pp 253–274.

35 Stumpf C, Rieger S, Fischer IU, Schierholz JM, Schietzel M: Vergleich der überlebenszeit bei patienten mit verschiedenen tumorentitäten – retrospektive untersuchung zur wirksamkeit von misteltherapie vs. daten eines tumorregisters; in: Scheer R, Alban S, Becker H, Holzgrabe U, Kemper FH, et al. (eds): Die Mistel in der Tumortherapie 2: Aktueller Stand der Forschung und Klinische Anwendung. Essen, KVC Verlag, 2009, pp 427–440.

36 Hager ED, Migeod F, Koomagi R, Schrittwieser G, Krautgartner I: Multimodale komplementäre therapie des fortgeschrittenen pankreaskarzinoms. Dtsch Zschr Onkol 2009;41:16–23.

37 Ritter PR, Tischoff I, Uhl W, Schmidt WE, Meier JJ: Sustained partial remission of metastatic pancreatic cancer following systemic chemotherapy with gemcitabine and oxaliplatin plus adjunctive treatment with mistletoe extract. Onkologie 2010;33:617–619.

38 Matthes H, Friedel WE, Bock PR, Zanker KS: Molecular mistletoe therapy: friend or foe in established anti-tumor protocols? A multicenter, controlled, retrospective pharmaco-epidemiological study in pancreas cancer. Curr Mol Med 2010;10:430–439.

39 Spahn G, Schindler V, Heyder C, Rieß H, Gerhards A, von Laue HB, Woernle M, Nabrotzki M, Leneweit G, Merkle A, Schad F: Clinical outcome study in pancreatic carcinoma using Viscum album therapy in an integrative approach. Phytomedicine 2011;18: 13.

40 Brandenberger M, Simoes-Wust AP, Rostock M, Rist L, Saller R: An exploratory study on the quality of life and individual coping of cancer patients during mistletoe therapy. Integr Cancer Ther 2012;11: 90–100.

41 Wiebelitz KR, Beer AM: Phytotherapy of chronic abdominal pain following pancreatic carcinoma surgery: a single case observation. Int J Gen Med 2012; 5:845–848.

42 Mansky PJ, Wallerstedt DB, Sannes TS, Stagl J, Johnson LL, Blackman MR, Grem JL, Swain SM, Monahan BP: NCCAM/NCI phase 1 study of mistletoe extract and gemcitabine in patients with advanced solid tumors. Evid Based Complement Alternat Med 2013;2013:964592.

43 Schad F, Axtner J, Buchwald D, Happe A, Popp S, Kroz M, Matthes H: Intratumoral mistletoe (Viscum album L) therapy in patients with unresectable pancreas carcinoma: a retrospective analysis. Integr Cancer Ther 2013;13:332–340.

44 Steele ML, Axtner J, Happe A, Kröz M, Matthes H, Schad F: Adverse drug reactions and expected effects to therapy with subcutaneous mistletoe extracts (Viscum album L.) in cancer patients. Evid Based Complement Alternat Med 2014;2014:724258.

45 Steele ML, Axtner J, Happe A, Kroz M, Matthes H, Schad F: Safety of intravenous application of mistletoe (Viscum album L.) preparations in oncology: an observational study. Evid Based Complement Alternat Med 2014;2014:236310.

46 Tröger W, Galun D, Reif M, Schumann A, Stankovic N, Milicevic M: Quality of life of patients with advanced pancreatic cancer during treatment with mistletoe: a randomized controlled trial. Dtsch Arztebl International 2014;111:493–502.

Asst. Prof. Danijel Galun
University Clinic for Digestive Surgery
Clinical Center of Serbia
Koste Todorovica 6, YU–11 000 Belgrade (Serbia)
E-Mail galun95@gmail.com

Zänker KS, Kaveri SV (eds): Mistletoe: From Mythology to Evidence-Based Medicine.
Transl Res Biomed. Basel, Karger, 2015, vol 4, pp 67–73 (DOI: 10.1159/000375427)

Dissecting the Anti-Inflammatory Effects of *Viscum album*: Inhibition of Cytokine-Induced Expression of Cyclo-Oxygenase-2 and Secretion of Prostaglandin E2

Sri Ramulu Elluru[a] · Chaitrali Saha[b–d] · Pushpa Hedge[b–d] ·
Alain Friboulet[e] · Jagadeesh Bayry[b, d, f, g] · Srini V. Kaveri[b, d, f, g]

[a]Department of Clinical and Experimental Medicine, Linköping University, Linköping, Sweden; [b]Institut
National de la Santé et de la Recherche Médicale, Unité 1138, Paris, [c]Université de Technologie
de Compiègne, Compiègne, [d]Centre de Recherche des Cordeliers, Equipe-Immunopathology and
Therapeutic Immunointervention, Paris, [e]CNRS UMR 6022, Université de Technologie de Compiègne,
Compiègne, [f]Sorbonne Universités, UPMC Univ Paris 06, UMRS 1138, Paris, [g]Université Paris Descartes,
Sorbonne Paris Cité, UMRS 1138, Paris, France

Abstract

Reports unraveling the beneficial effects of *Viscum album* (VA) preparations as complementary therapies for cancer have increasingly revealed the underlying molecular and cellular mechanisms, which encompass cytotoxic properties, induction of apoptosis, inhibition of angiogenesis and several other immunomodulatory mechanisms. In addition to their propitious relevance to cancer therapy, VA preparations are also relevant for the treatment of several inflammatory pathologies. In view of the intricate association of inflammation and cancer and the fact that several anti-tumor phytotherapeutics exert potent anti-inflammatory effects, we believe that an anti-inflammatory effect is responsible for the therapeutic benefits of VA preparations. One of the underlying molecular mechanisms of this inflammatory response is the selective down-regulation of the cyclo-oxygenase (COX)-2-mediated cytokine-induced secretion of prostaglandin E2 (PGE2). This inhibitory action has been associated with reduced expression of COX-2, without modulating COX-1 expression. This mechanism is associated with VA-induced destabilization of COX-2 mRNA, thereby depleting the functional COX-2 mRNA available for protein synthesis and for subsequent induction of secretion of PGE2. Together, these results demonstrate a novel anti-inflammatory mechanism of action of VA preparation, wherein VAQUSpez an anti-inflammatory effect by inhibiting cytokine-induced PGE2 via selective inhibition of COX-2 and destabilization of COX-2 mRNA.

Viscum album Preparations

Viscum album (VA) is extensively used as a complementary therapy for cancer and also in the treatment of several inflammatory pathologies [1]. VA preparations are standardized aqueous extracts of *Viscum album L.* (commonly known as European mistletoe, a semi-parasite that grows on different host trees), composed of mainly mistletoe lectins and viscotoxins [2, 3] and several other biologically active molecules, like flavonoids, several enzymes, peptides (such as viscumamide), amino acids, thiols, amines, polysaccharides, cyclitoles, lipids, phytosterols, triterpenes, phenylpropanes and minerals [4]. Currently, at least three VA preparations based on the host tree and the method of extraction are available for therapeutic application, including VA Qu Spez (oak tree), VA P (pini) and VA M (mali).

VA preparations have been used as a complementary therapy for several types of cancer, mainly in Europe and also, to some extent, in other parts of the world [1, 5]. The therapeutic benefits of VA preparations when utilized along with surgery, chemotherapy or radiotherapy contribute to the overall improvement of the quality of life of cancer patients [6, 7]. In addition, VA preparations have been implicated as conventional phytotherapeutics in the treatment of several conditions associated with nervous system abnormalities, allergic reactions, and immuno-inflammatory disorders [8–10].

The therapeutic benefits of VA preparations in diverse pathologies have been attributed to the method of preparation, the proportion of various bioactive compounds present within the extract and the host tree. In spite of its extensive use, the precise mechanisms associated with the anti-tumoral effects of VA are not yet clear. Accumulating evidence has revealed that these preparations exert anti-tumor activities that involve cytotoxic properties [11, 12], induction of apoptosis [13], inhibition of angiogenesis [14] and several other immunomodulatory and anti-inflammatory mechanisms [15, 16]. These properties collectively define the mechanistic basis for the therapeutic benefit of VA preparations, providing strong support for their application as complementary therapies in cancer. VA preparations also exert several immunostimulatory activities by interacting with the cellular and humoral compartments of the immune system, resulting in potent anti-tumor immune responses [16–19].

In view of the therapeutic benefits of VA preparations in diverse pathological situations, including inflammation and cancer, dissecting their molecular mechanisms would contribute enormously to the understanding of the roles of phytotherapy-based treatment strategies either in complementary and alternative medicine or in other combinational therapies. The modes of action of VA preparations underlying their therapeutic benefits in inflammatory pathologies are yet to be explored. However, the successful utilization of these preparations in the treatment of certain inflammatory pathologies raises several questions related to their mechanisms of action. In this chapter, we will discuss the anti-inflammatory mechanisms of VA, with a focus on inflammatory pathways.

Inflammation and Cyclo-Oxygenase-2

Inflammation is a physiopathological symptom of infection, autoimmunity or cancer. It is a basal physiological phenomenon that comprises a complex set of responses to tissue injury or to an infectious agent to eliminate the causative agent and to initiate the healing process. The interactions of innate and adaptive immune cells as well as nonimmune cells, such as endothelial cells and fibroblasts, with inflammatory stimuli induce the production of several molecular mediators [20–22]. These inflammatory mediators act on various target tissues and exert changes in the tissues' homeostatic functions. Thus, inflammation has to be regulated, which can be achieved by various anti-inflammatory agents, such as steroids, nonsteroidal anti-inflammatory agents (NSAIDs) [23], intravenous immunoglobulins [24], immunosuppressor cells [25], and neutralizing monoclonal antibodies to inflammatory cytokines [26].

Cyclo-oxygenases (COXs) are the regulatory enzymes of the prostaglandin E2 (PGE2) biosynthetic pathway that catalyze the rate-limiting step. They convert the free arachidonic acid in the cellular cytosol obtained upon degradation of membrane phospholipids into prostaglandin H2, an active precursor for the synthesis of various prostanoids. Among the isoforms of COXs, COX-1 is constitutively expressed in cells, whereas COX-2 is induced in response to inflammatory stimuli and significantly contributes to the induction of PGE2 [27]. PGE2 is a molecular mediator of several homeostatic functions, including those of the gastric mucosa and vascular endothelium [28]. However, it also exerts potent pro-inflammatory effects, such as the induction of fever and pain. Overproduction of PGE2 occurs in response to pro-inflammatory stimuli and correlates with the severity of certain infectious and inflammatory conditions [29, 30]. PGE2 exerts autocrine and paracrine actions on target cells and can induce pro-inflammatory reactions.

The expression pattern of COX-1 and COX-2 further regulate their differential functions. COX-1 is constitutively and stably expressed at low levels in many tissues, ensuring the constant production of prostaglandins, which are essentially required for the maintenance of important physiological functions, such as platelet aggregation, normal renal functions and gastric mucosal protection. In contrast, COX-2 is typically quiescent, but its expression can be induced in response to diverse pro-inflammatory and pathogenic stimuli. When stimulated, its expression is high and transient, which leads to a burst of prostaglandin production in a regulated and time-limited manner. Thus, depending on the COX isoform, the production of the same precursor prostaglandin H2 from arachidonic acid differs with respect to the amount and timing of production. This activity can be differentially decoded by the cells, thereby leading to the activation of various intracellular pathways involving specific classes of prostaglandins and therefore different responses [31].

Because COX-2 expression is up-regulated in several pathological conditions and human malignancies, strategies for controlling the expression and activity of COX-2 have been developed as potent anti-tumor and anti-inflammatory treatments [32]. In line with the therapeutic benefits of NSAIDs, which are synthetically designed mainly to inhibit the enzymatic activity of COX-2, a diverse spectrum of therapeutics of natural origin, such as phytotherapeutics, has been characterized to evaluate their potential to inhibit COX-2 functioning, thereby down-regulating the pathological level of prostaglandins. Due to the structural homology of COX-1 and COX-2, most NSAIDs inhibit both of these enzymes and therefore result in several severe side effects due to the inhibition of physiological prostaglandins. Therefore, selective inhibitors of COX-2 are of great interest. Although a promising class of synthetic COX-2-selective inhibitors called coxibs has been developed, their therapeutic efficacies are compromised due to various side effects [33]. Interestingly, clinical studies have revealed the selectivity of certain plant-derived molecules in inhibiting COX-2 that are as efficient as synthetic COX-2-specific antagonists (rofecoxib and celecoxib) in ameliorating both acute and chronic inflammatory conditions [34, 35].

Role of *Viscum album* in Inhibition of Cyclo-Oxygenase-2

The long-term side effects of NSAIDs in various pathological conditions and the increasing body of evidence demonstrating the anti-inflammatory activity of plant-derived molecules together encourage the concept of the use of phytotherapeutics as potent alternatives to classical anti-inflammatory drugs [34, 36]. With growing interest in promising new-generation anti-inflammatory therapeutics, studies exploring and characterizing novel phytotherapeutics with strong selectivity for COX-2 are of great value.

With the aim of understanding the role of VA in modulating the immuno-inflammatory response, our group analyzed the effects of VA preparation on the PGE2 axis and its regulation at the level of COX [37, 38]. Our results demonstrated that one of the anti-inflammatory functions of VA occurs through the inhibition of IL-1β-induced PGE2 biosynthesis. The expression pattern of cytokine-induced COX-2 detected in the presence of VA preparations confirmed that IL-1β induces the expression of COX-2 mRNA, while VA does not. In contrast, VA significantly inhibited the COX-2 protein expression induced by IL-1β. These results suggest that VA exerts a post-transcriptional regulatory effect on COX-2 expression [39].

However, the effect of VA preparation is not solely restricted to IL-1β-induced COX-2 expression. VA could also inhibit IFN-γ- and TNF-α-induced COX-2 expression (unpublished data), indicating that the suppression of COX-2 by VA occurred in response to inflammation induced by a wide range of cytokines. The fact that at the later phases of cytokine induction, VA did not inhibit COX-2 suggests that inhibition of COX-2 by VA occurs in the early phase of COX-2 regulation, but not in the later phases. Further, at each time point, the treatment of cells with VA resulted in reduced expression of IL-1β induced COX-2.

There was no significant difference in the degradation pattern of COX-2 in cytokine-stimulated cells with or without VA treatment. Additionally, VA induced destabilization of COX-2 mRNA, thereby diminishing the functional mRNA available for protein synthesis and for subsequent induction of secretion of PGE2 [38]. Thus, the modulation of COX-2 by VA strongly supports its value as an anti-inflammatory therapeutic.

Concluding Remarks

The prolonged administration of anti-inflammatory COX-2 inhibitors has been ineffectual for chemopreventive and chemotherapeutic purposes since the risks prevail over the benefits. The clinical demonstration of severe side effects due to the failure of classical COX-2 inhibitors to discriminate between aberrant pathological and homeostatic functional activation states has raised the concern that direct COX-2 enzymatic inhibition might not represent a sufficiently appropriate clinical strategy to target COX-2. Several phytotherapeutics have been shown to exert modulatory effects on COX-2 at various levels of its molecular regulation and therefore have been considered as effective alternative strategies for controlling the pathogenic expression of COX-2.

Given that VA preparation exerts potent anti-inflammatory effects by selective down-regulation of COX-2, it is extremely interesting to dissect the COX-2 inhibition mediated by VA in association with different regulatory mechanisms at the molecular level. VA exerts anti-inflammatory effects by interfering with cytokine-induced PGE2 biosynthesis through selective inhibition of the COX-2 protein, suggesting a beneficial role with minimal side effects. These observations are relevant to understand the mechanisms of action of VA preparations and may provide insights for the further exploration of their anti-inflammatory mechanisms in diverse pathologies. Interestingly, in these studies, no changes in the expression of COX-1 were observed at any of the concentrations of VA preparation that inhibited COX-2, irrespective of robust stimulation by IL-1β. These findings suggest a strong selectivity in the anti-inflammatory mechanisms of VA preparation in inhibiting COX-2 expression. Given that VA exerts potent anti-inflammatory effects by the selective down-regulation of COX-2, it is extremely interesting to dissect the COX-2 inhibition mediated by VA in association with different regulatory mechanisms at the molecular level. Increasing evidence for the anti-inflammatory mechanisms of action of these preparations will lead to greater acceptance of their use for the treatment of various pathologies.

Acknowledgments

Our work is supported by the Institut National de la Santé et de la Recherche Médicale, the Centre National de la Recherche Scientifique, the Regional Program Bio-Asie 2010 by the French Ministry of Foreign and the European Affairs and Institut Hiscia, Arlesheim, Switzerland.

References

1 Bock PR, Friedel WE, Hanisch J, Karasmann M, Schneider B: [Efficacy and safety of long-term complementary treatment with standardized European mistletoe extract (Viscum album L.) in addition to the conventional adjuvant oncologic therapy in patients with primary non-metastasized mammary carcinoma. Results of a multi-center, comparative, epidemiological cohort study in Germany and Switzerland]. Arzneimittelforschung 2004;54:456–466.

2 Franz H, Ziska P, Kindt A: Isolation and properties of three lectins from mistletoe (Viscum album L.). Biochem J 1981;195:481–484.

3 Olsnes S, Stirpe F, Sandvig K, Pihl A: Isolation and characterization of viscumin, a toxic lectin from Viscum album L. (mistletoe). J Biol Chem 1982;257:13263–13270.

4 Urech K, Schaller G, Jaggy C: Viscotoxins, mistletoe lectins and their isoforms in mistletoe (Viscum album L.) extracts Iscador. Arzneimittelforschung 2006;56:428–434.

5 Kaegi E: Unconventional therapies for cancer: 3. Iscador. Task Force on Alternative Therapies of the Canadian Breast Cancer Research Initiative. CMAJ 1998;158:1157–1159.

6 Bussing A, Schietzel M: Apoptosis-inducing properties of Viscum album L. extracts from different host trees, correlate with their content of toxic mistletoe lectins. Anticancer Res 1999;19:23–28.

7 Kienle GS, Kiene H: Review article: Influence of Viscum album L (European mistletoe) extracts on quality of life in cancer patients: a systematic review of controlled clinical studies. Integr Cancer Ther 2010;9:142–157.

8 Christen-Clottu O, Klocke P, Burger D, Straub R, Gerber V: Treatment of clinically diagnosed equine sarcoid with a mistletoe extract (Viscum album austriacus). J Vet Intern Med 2010;24:1483–1489.

9 Radenkovic M, Ivetic V, Popovic M, Mimica-Dukic N, Veljkovic S: Neurophysiological effects of mistletoe (Viscum album L.) on isolated rat intestines. Phytother Res 2006;20:374–377.

10 Tusenius KJ, Spoek AM, van Hattum J: Exploratory study on the effects of treatment with two mistletoe preparations on chronic hepatitis C. Arzneimittelforschung 2005;55:749–753.

11 Duong Van Huyen JP, Delignat S, Kazatchkine MD, Kaveri SV: Comparative study of the sensitivity of lymphoblastoid and transformed monocytic cell lines to the cytotoxic effects of Viscum album extracts of different origin. Chemotherapy 2003;49:298–302.

12 Duong Van Huyen JP, Sooryanarayana, Delignat S, Bloch MF, Kazatchkine MD, Kaveri SV: Variable sensitivity of lymphoblastoid cells to apoptosis induced by Viscum album Qu FrF, a therapeutic preparation of mistletoe lectin. Chemotherapy 2001;47:366–376.

13 Duong Van Huyen JP, Bayry J, Delignat S, Gaston AT, Michel O, Bruneval P, Kazatchkine MD, Nicoletti A, Kaveri SV: Induction of apoptosis of endothelial cells by Viscum album: a role for anti-tumoral properties of mistletoe lectins. Mol Med 2002;8:600–606.

14 Elluru SR, Duong Van Huyen JP, Delignat S, Prost F, Heudes D, Kazatchkine MD, Friboulet A, Kaveri SV: Antiangiogenic properties of viscum album extracts are associated with endothelial cytotoxicity. Anticancer Res 2009;29:2945–2950.

15 Elluru S, Duong Van Huyen JP, Delignat S, Prost F, Bayry J, Kazatchkine MD, Kaveri SV: Molecular mechanisms underlying the immunomodulatory effects of mistletoe (Viscum album L.) extracts Iscador. Arzneimittelforschung 2006;56:461–466.

16 Elluru SR, Duong van Huyen JP, Delignat S, Kazatchkine MD, Friboulet A, Kaveri SV, Bayry J: Induction of maturation and activation of human dendritic cells: a mechanism underlying the beneficial effect of Viscum album as complimentary therapy in cancer. BMC Cancer 2008;8:161.

17 Duong Van Huyen JP, Delignat D, Bayry J, Kazatchkine MD, Bruneval P, Nicoletti A, Kaveri SV: Interleukin-12 is associated with the in vivo anti-tumor effect of mistletoe extracts in B16 mouse melanoma. Cancer Lett 2006;243:32–37.

18 Heinzerling L, von Baehr V, Liebenthal C, von Baehr R, Volk HD: Immunologic effector mechanisms of a standardized mistletoe extract on the function of human monocytes and lymphocytes in vitro, ex vivo, and in vivo. J Clin Immunol 2006;26:347–359.

19 Hostanska K, Hajto T, Spagnoli GC, Fischer J, Lentzen H, Herrmann R: A plant lectin derived from Viscum album induces cytokine gene expression and protein production in cultures of human peripheral blood mononuclear cells. Nat Immun 1995;14:295–304.

20 Aimanianda V, Haensler J, Lacroix-Desmazes S, Kaveri SV, Bayry J: Novel cellular and molecular mechanisms of induction of immune responses by aluminum adjuvants. Trends Pharmacol Sci 2009;30:287–295.

21 Bayry J: Immunology: TL1A in the inflammatory network in autoimmune diseases. Nat Rev Rheumatol 2010;6:67–68.

22 Medzhitov R: Inflammation 2010: new adventures of an old flame. Cell 2010;140:771–776.

23 Achoui M, Appleton D, Abdulla MA, Awang K, Mohd MA, Mustafa MR: In vitro and in vivo anti-inflammatory activity of 17-O-acetylacuminolide through the inhibition of cytokines, NF-kappaB translocation and IKKbeta activity. PLoS One 2010; 5:e15105.

24 Bayry J, Negi VS, Kaveri SV: Intravenous immunoglobulin therapy in rheumatic diseases. Nat Rev Rheumatol 2011;7:349–359.

25 Bayry J, Siberil S, Triebel F, Tough DF, Kaveri SV: Rescuing CD4+CD25+ regulatory T-cell functions in rheumatoid arthritis by cytokine-targeted monoclonal antibody therapy. Drug Discov Today 2007; 12:548–552.

26 Tomlinson KL, Davies GC, Sutton DJ, Palframan RT: Neutralisation of interleukin-13 in mice prevents airway pathology caused by chronic exposure to house dust mite. PLoS One, 2010;5:e13136.

27 Pairet M, Engelhardt G: Distinct isoforms (COX-1 and COX-2) of cyclooxygenase: possible physiological and therapeutic implications. Fundam Clin Pharmacol 1996;10:1–17.

28 Pozzi A, Zent R: Regulation of endothelial cell functions by basement membrane- and arachidonic acid-derived products. Wiley Interdiscip Rev Syst Biol Med 2009;1:254–272.

29 Passwell J, Levanon M, Davidsohn J, Ramot B: Monocyte PGE2 secretion in Hodgkin's disease and its relation to decreased cellular immunity. Clin Exp Immunol 1983;51:61–68.

30 Redondo S, Ruiz E, Gordillo-Moscoso A, Navarro-Dorado J, Ramajo M, Rodriguez E, Reguillo F, Carnero M, Casado M, Tejerina T: Overproduction of cyclo-oxygenase-2 (COX-2) is involved in the resistance to apoptosis in vascular smooth muscle cells from diabetic patients: a link between inflammation and apoptosis. Diabetologia 2011;54:190–199.

31 Kam PC, See AU: Cyclo-oxygenase isoenzymes: physiological and pharmacological role. Anaesthesia 2000;55:442–449.

32 Martel-Pelletier J, Pelletier JP, Fahmi H: Cyclooxygenase-2 and prostaglandins in articular tissues. Semin Arthritis Rheum 2003;33:155–167.

33 Marnett LJ: The COXIB experience: a look in the rearview mirror. Annu Rev Pharmacol Toxicol 2009; 49:265–290.

34 Chrubasik S, Kunzel O, Model A, Conradt C, Black A: Treatment of low back pain with a herbal or synthetic anti-rheumatic: a randomized controlled study. Willow bark extract for low back pain. Rheumatology (Oxford) 2001;40:1388–1393.

35 Cravotto G, Boffa L, Genzini L, Garella D: Phytotherapeutics: an evaluation of the potential of 1,000 plants. J Clin Pharm Ther 2010;35:11–48.

36 Meyer-Kirchrath J, Schror K: Cyclooxygenase-2 inhibition and side-effects of non-steroidal anti-inflammatory drugs in the gastrointestinal tract. Curr Med Chem 2000;7:1121–1129.

37 Hegde P, Maddur MS, Friboulet A, Bayry J, Kaveri SV: Viscum album exerts anti-inflammatory effect by selectively inhibiting cytokine-induced expression of cyclooxygenase-2. PLoS One 2011;6:e26312.

38 Saha C, Hegde P, Friboulet A, Bayry J, Kaveri SV: Viscum album-mediated COX-2 inhibition implicates destabilization of COX-2 mRNA. PLoS One 2015;10:e0114965.

39 Tetsuka T, Baier LD, Morrison AR: Antioxidants inhibit interleukin-1-induced cyclooxygenase and nitric-oxide synthase expression in rat mesangial cells. Evidence for post-transcriptional regulation. J Biol Chem 1996;271:11689–11693.

Srini V. Kaveri
INSERM Unité 1138, Equipe-Immunopathology and
Therapeutic Immunointervention
Centre de Recherche des Cordeliers
15 rue de l'Ecole de Médicine, Paris, FR–75006 (France)
E-Mail srini.kaveri@crc.jussieu.fr

Zänker KS, Kaveri SV (eds): Mistletoe: From Mythology to Evidence-Based Medicine.
Transl Res Biomed. Basel, Karger, 2015, vol 4, pp 74–81 (DOI: 10.1159/000375428)

Dancing with the Devil: Cancer-Related Fatigue, from Inflammation to Treatment Options

Kurt S. Zänker

Institute of Immunology and Experimental Oncology, Centre for Biomedical Education and Research,
Department of Human Medicine, Faculty of Health Science, University Witten/Herdecke, Witten, Germany

Abstract

Cancer-related fatigue (CRF) is the most common debilitating symptom in cancer, affecting almost 90% of cancer patients. CRF might persist for years after treatment. CRF is not relieved by rest or sleep, as opposed to a 'burnout syndrome' in otherwise healthy persons. CRF is a disease entity by itself. Clinical studies have identified genetic, biological, psychosocial, and behavioral risk factors associated with CRF. Currently, there are limited intervention options to effectively treat CRF. Co-morbidities, as an attribute of increasing age and as contributors to CRF, should be diagnosed and treated either by pharmacological or by nonpharmacological interventions, similar to CRF. Chronic (neuro-)inflammation and chaotic (neuro-)immune signaling are up-coming biomarkers and targets for personalized treatment. Because of their proven anti-inflammatory properties, herbal remedies (Wisconsin ginseng, mistletoe extracts) have been used successfully in clinical trials to treat CRF. There are almost no adverse effects caused or amplified in combination with anti-cancer drugs and/ or radiotherapy when the indispensable conventional first- or second-line therapies were supported (supportive care) by well-defined extracts from the plant kingdom. Patients currently suffering from CRF cannot wait for the 'magic bullet', so for the time being, those observational results should be considered in clinical chemo- or chemo-radiotherapy protocols to fight against CRF and to increase quality of life.
© 2015 S. Karger AG, Basel

Introduction

Cancer-related fatigue (CRF) is a substantial symptom burden that often affects a patient chronically along the trajectory of the disease. The National Comprehensive Cancer Network describes CRF as distressing and persistent [1]. CRF is a multidimensional symptom experienced physically, cognitively and emotionally, with a prevalence of 60–90%, depending on the patient population, cancer treatment protocols for first- and second-line treatment and the method of assessment. CRF is characterized by feelings of tiredness, weakness and lack of energy and is distinct from 'normal' drowsiness experienced by healthy individuals because CRF is not relieved by rest or sleep [2]. The effect of CRF is as severe as cancer itself because it reduces the patient's willingness to adhere to treatment and might therefore be a risk factor for reduced survival. The symptoms are often underreported by patients and underestimated by the physician and are therefore scarcely treated.

Symptoms of Cancer-Related Fatigue

Nausea and vomiting, bleeding, diarrhea and constipation are some of the etiological factors that can be pharmacologically addressed in a supportive care intervention. Neurological disorders, like impairment of memory and associative memory, distracted attention, worrying, nervousness and nervous irritability, weakness, tremors and impaired muscle function, have a negative impact on patients' overall well-being and because of its complexity are often leading to inappropriate treatment. Lack of appetite, not being able to get things done, pain and disturbed sleep decrease the level of the much-needed energy of the affected patients. Depression and anxiety are often associated with CRF (mental fatigue) [3]. All of these entities strongly and profoundly interfere with day-to-day quality of life and can even persist for years after treatment.

Mechanisms of Cancer-Related Fatigue

CRF is a multifactorial symptom cluster and is presumably caused and modulated by biological, psychosocial, emotional and behavioral factors, including contextual factors like co-morbidity, nutritional issues, smoking and alcohol consumption. At the very beginning of CRF research, therapy-induced bleeding and anemia were believed to be the main factors. Platelet transfusion and recombinant erythropoietic agents were given to stop bleeding and to increase hemoglobin levels. However, several major concerns have been raised regarding potential risks associated with the use of recombinant erythropoietins, including thrombotic events, red-cell aplasia, and growth stimulation of certain tumor types. Chaotic cytokine storms and hypothalamic-pituitary-adrenal (HPA) axis and central and peripheral neurotransmitter dysregulation, caused by diabetes as a co-morbidity, have gained the most empirical attention. In fact, the most attractive approach focuses on pro-inflammatory cytokines and the persistence of chronic inflammation [4, 5]. Tumor growth increases neuro-inflammation and depressive-like behavior and leads to alteration in muscle function according to a mouse model that can discriminate between loss of muscle function and altered mood/motivation [6]. Chemo- and radiotherapy can activate the pro-inflammatory

pharmacological agents for CRF. Psychostimulants should be used only in patients with moderate to severe CRF and with advanced cancer or in those receiving active cancer therapy. There is no evidence regarding psychostimulant use in cancer patients who have completed active cancer therapy [27]. Existing trials of methylphenidate provide limited evidence for its use to fight against CRF [28]. Cyproheptadine, hydrazine, metoclopramide, pentoxifylline, megestrol acetate, medroxyprogesterone acetate, oxandrolone, omega-3 fatty acids, cannabinoids, bortezomib, ghrelin and melanocortin antagonists exhibited a strong rationale for use but failed or showed equivocal results in clinical trials. A new era for anti-CRF drugs is arising on the horizon, consisting of agents targeting inflammation and pro-inflammatory cytokine activity.

Use of Herbal Medicine to Improve Cancer-Related Fatigue

Herbal remedies are part of traditional and folk healing methods with long histories of use. Unlike Western medicine, which generally uses single purified or synthesized compounds and aims to target a specific receptor or pathway molecule, herbal extracts based on ethno-pharmacological knowledge usually comprise multiple either well- or ill-defined compounds that are necessary for efficacy. There is a plethora of publications showing an anti-cancer effect in vitro and in animal models; however, meta-analyses of clinical trials suggest potential but mostly do not confirm a substantial therapeutic effect according to Western standards for clinical trials. The backbone of current cancer treatment remains those agents that have been rigorously proven to have clinical anti-cancer activity alone or in combination. However, in order to ameliorate the side effects of conventional cancer therapies, herbal extracts, which do not cause adverse effects by themselves, provide a strong rationale for combined-modality therapies [29, 30]. Clinical data support the benefit of American ginseng in improving CRF [31]. In retrospective clinical studies (pancreatic and colorectal cancer), we tested the hypothesis that the anti-inflammatory activity of mistletoe extract (ISCADOR®) would lead to decreased CRF symptoms because in vitro results strongly suggested anti-inflammatory gene expression [32] and inhibition of the cytokine-induced pro-inflammatory activity of COX-2 [33]; both molecular entities were significantly reduced by mistletoe extracts.

Supportive Care for and Treatment of Cancer-Related Fatigue with Mistletoe Preparations

Mistletoe preparations (Viscum album L. (VaL)) are used for supportive care in cancer treatment. It is well documented that mistletoe preparations have cytotoxic properties in vitro, induce apoptosis, modulate immune responses and show tumor vasculature-disrupting activity. However, the use of mistletoe preparations in the clinical setting as a first- or second-line therapy is highly debated. Very recently, Tröger et al. [34] published the results of a prospective, parallel, open-label, monocenter, group-sequential, randomized phase III study. Patients with locally advanced or metastatic pancreatic cancer received escalating doses of 0.01 mg up to 10 mg VaL extract. The results suggested that VaL is a nontoxic and effective second-line therapy for patients with locally advanced or metastatic pancreatic cancer. The median overall survival was 6.6 months within a 'good prognosis group' (versus 3.3 months in the control group) and 3.4 months within a 'poor prognosis group' (versus 2.0 months in the control group) [34]. Moreover, it was reported that disease-related symptoms, including CRF, were alleviated. These supportive care results are in accordance with retrospectively obtained clinical results addressing CRF in pancreatic [30] and colorectal cancer [35] patients. The retrospective assessment of CRF in pancreatic and colorectal cancer patients strongly supported VaL medication within chemo- and chemo-radiotherapy protocols in identified and diagnosed patients with CRF, with the intention to deliver supportive care that palliates CRF and improves quality of life.

Conclusion

In general medicine, chronic fatigue syndrome (CFS) is a poorly understood condition affecting a broad range of physical and mental systems in the body of many people. In oncology, the etiology of CRF is also not yet clear, but experimental and clinical results provide support for further evaluating hypothesis-driven questions about the role of inflammatory processes in CRF. CRF is not relieved by rest or sleep and is associated with other medical conditions (co-morbidities). This is in contrast to CFS in otherwise-healthy persons, who can overcome CFS by resting, sleeping and lowering their workload. Currently, there is no gold standard to treat CRF, but we and others have shown that VaL is a promising candidate for inclusion into conventional anti-tumor protocols. Presently, the anti-inflammatory property of VaL, e.g. selective inhibition of cytokine-induced expression of COX-2, is a target for personalized intervention to reduce the burden of CRF symptoms and improve quality of life [36].

Acknowledgment

KSZ is supported by the Fritz-Bender-Foundation, Munich, Germany.

References

1 Piper BF, Cella D: Cancer-related fatigue and clinical subtypes. J Natl Compr Canc Netw 2010;8:958–966.
2 Hofmann M, Ryan JL, Figueroa-Moseley CD, Jean-Pierre P, Morrow GR: Cancer-related fatigue: the scale of the problem. Oncologist 2007;12(suppl 1):4–10.
3 Brown LF, Kroenke K: Cancer-related fatigue and its association with depression and anxiety: a systemic review. Psychosomatics 2009;50:440–447.
4 de Raaf PJ, Sleijfer S, Lamers CH, Jager A, Gratama JW, van der Rijt CC: Inflammation and fatigue dimensions in advanced cancer patients and cancer survivors: an explorative study. Cancer 2012;118: 6005–6011.
5 Bower JE, Lamkin DM: Inflammation and cancer-related fatigue: mechanisms, contributing factors and treatment implications. Brain Behav Immun 2013;30(suppl):S48–S57.
6 Norden DM, Bicer S, Clark Y, Jing R, Henry CJ, Wold LE, Reiser PJ, Godbout JP, McCarthy DO: Tumor growth increases neuroinflammation, fatigue and depressive-like behavior prior to alterations in muscle function. Brain Behav Immun 2015;43:76–85.
7 Yennurajalingam S, Bruera E: Role of corticosteroids for fatigue in advanced incurable cancer: is it a 'wonder drug' or 'deal with the devil'. Curr Opin Support Palliat Care 2014;8:346–351.
8 Morgan N, Irwin MR, Chung M, Wang C: The effects of mind-body therapies in the immune system: meta-analysis. 2014;9:e100903.
9 Dearlove J: Putting humpty together again: Homo sociologicus, homo economicus and the politcal science of the British state. Europ J Political Sci 1995;27: 477–505.
10 Jaremka LM, Fagundes CP, Glaser R, Bennett JM, Malarkey WB, Kiecolt-Glaser JK: Loneliness predicts pain, depression, and fatigue: understanding the role of immune dysregulation. Psychoneuroendocrinology 2013;38:1310–1317 .
11 Yennurajalingam S, Bruera E: Review of clinical trials of pharmacologic interventions for cancer-related fatigue: focus on psychostimulants and steroids. Cancer J 2014;20:319–324.
12 Niggemann B, Zänker KS: Immunologische Grundlagen der Psychoneuroimmunlogie; in Schubert C (ed): Psychoneuroimmunologie und Psychotherapie. Stuttgart, Schattauer, 2011.
13 Harris B, Ross J, Sanchez Reilly S: Sleeping in the arms of cancer: a review of sleeping disorders among patients with cancer. Cancer J 2014;20:299–305.
14 Pachmann DR, Price KA, Carey EC: Nonpharmacological approach to fatigue in patients with cancer. Cancer J 2014;20:313–318.
15 Dhillon HM, van der Ploeg HP, Bell ML, Boyer M, Clarke S, Vardy J: The impact of physical activity on fatigue and quality of life in lung cancer patients: a randomised controlled trial protocol. BMC Cancer 2012;12:572.
16 Steindorf K, Schmidt ME, Klassen O, Ulrich CM, Oelmann J, Habermann N, Beckhove N, Owen R, Debus J, Wiskemann J, Potthoff K: Randomized, controlled trial of resistance training in breast cancer patients receiving adjuvant radiotherapy: results on cancer-related fatigue and quality of life. Ann Oncol 2014;25:2237–2243.
17 Yang TY, Chen ML, Li CC: Effects of an aerobic exercise programme on fatigue for patients with breast cancer undergoing radiotherapy. J Clin Nurs 2014, DOI: 101111/jocn 12672 Epub ahead of print.
18 Tang WR, Chen Wj, Yu CT, Cahng YC, Chen CM, Wang CH, Yang SH: Effects of acupressure on fatigue of lung cancer patients undergoing chemotherapy: an experimental pilot study. Complement Ther Med 2014;22:581–591.
19 Cramer H, Lauche R, Paul A, Langhorst J, Kümmel S, Dobos GJ: Hypnosis in breast cancer care: a systematic review of randomized controlled trials. Integr Cancer Ther 2015;14:5–15 .
20 Bower JE: Cancer-related fatigue – mechanisms, risk factors, and treatments. Nat Rev Clin Oncol 2014;11: 597–609.
21 Mantovani C, Maccio A, Madeddu C, Cramignano C, Lusso MR, Serpe R, Massa E, Astara C, Deiana L: A phase II study with antioxidants, both in the diet and supplemented, pharmaconutritional support, progestagen, and anti-cyclooxygenase-2 showing efficacy and safety in patients with cancer-related anorexia/cachexia and oxidative stress. Cancer Epidemiol Biomarkers Prev 2006;15:1030–1034.
22 Zic SM, Sen A, Han-Markey TL, Harris RE: Examination of the association of diet and persistent cancer-related fatigue: a pilot study. Oncol Nurs Forum 2013;40:E41–E49.
23 Alfano CM, Day JM, Katz ML, Herndon JE 2nd, Billoni MA, Oliveri JM, Donohue K, Paskell ED: Exercise and dietary change after diagnosis and cancer-related symptoms in long-term survivors of breast cancer: CALGB 79804. Psychooncology 2009;18: 128–133.
24 Carayol M, Romieu G, Bleuse JP, Senesse P, Gourgou-Bourgade S, Sari C, Jacot W. Sancho-Garnier H, Janiszewsik C, Launay S, Cousson-Gelie F, Ninot G: Adapted physical activity and diet (APAD) during adjuvant breast cancer therapy: design and implementation of a prospective randomized controlled trial. Contemp Clin Trials 2013;36:531–543.

25 George SM, Alfano CM, Neuhouser ML, Smith AW, Baumgartner RN, Baumgartner KB, Bernstein L, Bailard-Sarbash R: Better postdiagnosis diet quality is associated with less cancer-related fatigue in breast cancer survivors. J Cancer Surviv 2014;8:680–687.

26 Denlinger CS, Ligibel JA, Are M, Baker KS, Demark-Wahnefried W, Friedman DL, Goldman M, Jones L, King A, Kur GH, Kvale E, Langbaum TS, Leondardi-Warren K, McCabe MS, Melisko M, Montoya JG, Mooney K, Morgan MA, Moslehi JJ, O'Connor T, Overholser L, Paskett ED, Raza M, Svriala KL, Urba SG, Wakabayashi MT, Zee P, McMillian N, Freedman-Cass D: Survivorship: fatigue, version 1.2014. J Natl Compr Canc Netw 2014;12:876–887.

27 Bower JE, Bak K, Berger A, Breitbart W, Escalante CP, Ganz PA, Schnipper HH, Lacchetti C, Ligibel JA, Lyman GH, Ogaily MS, Piri WF, Jacobsen PB; American Society of Clinical Oncology: Screening, assessment, and management of fatigue in adult survivors of cancer: an American Society of Clinical Oncology clinical practice guideline adaptation. J Clin Oncol 2014;32:1840–1850.

28 Gong S, Sheng P, Jin H, He H, Qi E, Chen VV, Dong Y, Hou L: Effect of methylphenidate in patients with cancer-related fatigue: a systematic review and meta-analysis. PLoS One 2014;9:e84391.

29 Friedel WE, Matthes H, Bock PR, Zaenker KS: Systemic evaluation of the clinical effects of supportive mistletoe treatment within chemo- and/or radio-therapy protocols and long-term mistletoe application in non-metastatic colo-rectal carcinoma: a multicenter, controlled, observational cohohrt study. J Soc Integr Oncol 2009;7:137–145.

30 Matthes H, Friedel WE, Bock PR, Zänker KS: Molecular mistletoe therapy: friend or foe in established anti-tumor protocols? A multicenter, controlled, retrospective pharmaco-epidemiological study in pancreas cancer. Curr Mol Med 2010;10:430–439.

31 Barton DL, Liu H, Dakhil SR, Linquist B, Soan JA, Nichols CR, McGinn TW, Stella PJ, Seeger GR, Sood A, Loprinzi CL: Wisconsin Ginseng (Panax quinquefolius) to improve cancer-related fatigue: a randomized, double-blind trial, N07C2. J Natl Cancer Inst 2013;105:1230–1238.

32 Wagschal I, Eggenschwiler J, von Balthazar L, Patrignani A, Rehrauer H, Schlapbach R, Ramos MH, Viviani A: Gene expression signatures of pathway alterations in tumor cells caused by plant extracts. Medicina (Buenos Aires) 2007;6:97–106.

33 Hegde P, Maddur MS, Friboulet A, Bayry J, Kaveri SV: Viscum album exerts anti-inflammatory effect by selectively inhibiting cytokine-induced expression of cyclooxygenase-2. PloS One 2011;6:e26312.

34 Tröger W, Galun D, Reif M, Schumann A, Stankovic N, Milicevic M: Viscum album (L.) extract therapy in patients with locally advanced or metastatic pancreatic cancer: a randomised clinical trial on overall survival. Eur J Cancer 2013;49:3788–3797.

35 Bock PR, Hanisch J, Matthes H, Zänker KS: Targeting inflammation in cancer-related fatigue: a rationale for mistletoe therapy as supportive care in colorectal cancer patients. Inflamm Allergy Drug Targerts 2014;13:105–111.

36 Zaenker KS: Cancer-related fatigue casts a dark shadow over the quality of life of cancer patients. Inflamm Cell Sign 2014;1:212–214.

Kurt S. Zänker, MD, DVM
Professor of Immunology and Experimental Oncology
Institute of Immunology and Experimental Oncology, University Witten/Herdecke
Stockumerstrassee 10, DE–58448 Witten (Germany)
E-Mail ksz@uni-wh.de

Author Index

Subject Index